# Scotland's Mountain Landscapes

Other earth science titles from Dunedin include:

*The Western Highlands of Scotland* (2019)
Classic Geology in Europe
Con Gillen
ISBN: 9781780460406

*Hutton's Arse* (2019)
Second edition
Malcolm Rider and Peter Harrison
ISBN: 9781780460932

*Volcanoes and the Making of Scotland* (2015)
Second cedition
Brian Upton
ISBN: 9781780460567

*Scottish Fossils* (2013)
Nigel Trewin
ISBN: 9781780460192

*Geology and landscapes of Scotland* (2013)
Con Gillen
ISBN: 9781780460093

For further details of these and other Dunedin
Earth and Environmental Sciences titles see
www.dunedinacademicpress.co.uk

# Scotland's Mountain Landscapes

## A GEOMORPHOLOGICAL PERSPECTIVE

### Colin K. Ballantyne

EDINBURGH ◆ LONDON

Published by Dunedin Academic Press Ltd
*Head Office*: Hudson House, 8 Albany Street, Edinburgh EH11 3QB
*London Office*: The Towers, 352 Cromwell Tower, Barbican, EC2Y 8NB

www.dunedinacademicpress.co.uk

ISBNs
9781780460796 (Hardback)
9781780466279 (PDF)
9781780466101 (ePub)
9781780466118 (Kindle)

*British Library Cataloguing in Publication data*
A catalogue record for this book is available from the British Library

Typeset by Makar Publishing Production, Edinburgh
Printed in Poland by Hussar Books

For Rebecca, Hamish and Kate

# Contents

# Acknowledgements

In preparing this book for submission I owe a huge debt of gratitude to Professor John Gordon. John has an unrivalled knowledge and understanding of the geomorphology of Scotland, and his insightful and incisive comments greatly improved the text. The book has also greatly benefited from the contribution of my longstanding friend and colleague Graeme Sandeman, cartographer at the University of St Andrews, who produced all the final maps and figures. Graeme has been transforming my inept scribbles into crisp cartography for nearly four decades, and the figures contained herein are (yet again) testimony to his skill and forbearance. I am also indebted to Professor Doug Benn and Dr Adrian Hall for reading over particular chapters and preventing a few embarrassing blunders. John Gordon, Charles Warren, David Evans and Martin Kirkbride all helped by contributing photographs.

The impetus for writing this book came from friends who have for many years had to suffer my enthusiasm for explaining the origin of ploughing boulders, tors, rock-slope deformations and a multitude of other landforms as we tramped over Scottish mountains. I thank in particular Peter Robinson for being a patient listener for over 40 years, and Cairns Dickson, who egged me on to put it all in print, and who gamely read through and commented on the draft manuscript. I owe an enormous debt to my co-researchers on aspects of Scotland's geomorphology, notably Professors Doug Benn, Svein Olaf Dahl, Atle Nesje, Danny McCarroll and John Stone, my former PhD students, and especially the legions of enthusiastic students at St Andrews University who have clambered up Scottish mountains in all weathers to assist in fieldwork.

I owe particular thanks to Anthony Kinahan of Dunedin Academic Press for providing encouragement and advice throughout the writing and production stages, David McLeod for design and compilation of the book, and Anne Morton for eagle-eyed copy-editing.

This book owes much to my mentor and friend, Dr Brian Sissons (1926–2018). Brian made an unparalleled contribution to our understanding of the geomorphology of Scotland, some of which is reflected here. His seminal book *The Evolution of Scotland's Scenery* (1967) influenced my decision to pursue a career as a geomorphologist, a choice that has allowed me to combine research with my love of mountains, the Arctic and above all the Scottish landscape.

My children, Hamish and Kate, have tolerated numerous 'holidays' that involved them being cajoled up rainswept mountains with promises of ice cream. My wife Rebecca has been my constant companion on the hills in weather fair and foul, and tolerated my long absences spent working in a hut at the bottom of the garden. Together we treasure wonderful memories of summits across the globe from Kilimanjaro to Kosciuzsco, but the best of these have been in Scotland.

Colin Ballantyne
Blebo Craigs
Scotland

# Chapter 1

# Introduction

*The lowly offices of wind and rain, springs and frost, snow and ice, trifling as they may appear, have nevertheless been chosen as instruments to carve the giant frame-work of the mountains.*
Archibald Geikie: *The Scenery of Scotland* (1865)

## The land of the mountain and the flood

Mountains represent the essence of Scotland's scenery. Postcards, calendars, shortbread tins and dishtowels seeking to capture 'Scotland' in a single image almost invariably depict a mountainous Highland landscape, sometimes with a loch or castle in the middle distance and a kilted bagpiper artfully (and sometimes digitally) inserted in the foreground. They are a fundamental part of Scottish identity, for though most Scots live in the towns and cities of the lowlands, the mountains are close by on the horizon, a line of purple or snow-covered peaks that reminds us of another Scotland where the interplay of sunshine and cloud over rocky crags, deep lochs, windswept plateaux and lonely moorlands creates a sense of wilderness and an opportunity to escape from our urban hinterland.

And escape we do. When the weather is favourable, thousands of hillwalkers visit the summits of Scotland's mountains every week, many with the aim of completing the ascent of the 282 Munros (summits over 3000 feet or 914 m). Rock climbers are drawn to the crags and corries of Skye and Glen Coe, skiers to the snowy slopes of Glen Shee and the Cairngorms, and others come to the mountains for fell running, mountain biking and even hang gliding. Outnumbering all of these, however, are the visitors who come simply to marvel at the most wonderful scenery in the British Isles.

Nobody who has explored the mountains of Scotland can fail to have been impressed by their sheer diversity. The isolated sandstone peaks of the far northwest, the serrated gabbro ridges of Skye, the granite high plateaux of the Cairngorms, and the rolling uplands of southern Scotland (Fig. 1.1) represent a variety of mountain landscapes that rivals any on Earth. Such diversity reflects not only Scotland's tumultuous geological evolution, which has created a mosaic of contrasting rock

types, but also the operation of a wide range of erosional processes that have sculpted the underlying rocks into the wonderful topographic variety of Scotland's mountain landscapes. Some of these processes operated in deep time, many millions of years ago, others throughout the Ice Age, and many, such as frost action, rockfall and river erosion, have continued to modify mountain landscapes since the disappearance of the last glaciers.

In this book we shall take a journey through time, beginning with the formation of the oldest rocks and ending with the manifold processes that are still operating on high ground. We shall visit past eras when Scotland lay near the Equator, when alpine-scale mountains towered over the landscape, when volcanoes spewed out copious lava flows, when ice covered the land, and when earthquakes triggered major landslides. Mountain scenery is always inspirational; but understanding how mountain landscapes have evolved deepens our perspectives of time and space, and the transience of human existence. Just 12,500 years ago, for example, Scotland looked like high-arctic Svalbard today: a great icefield occupied the western Highlands, a valley glacier was advancing to the southern end of Loch Lomond, permafrost underlay the ground beyond the glaciers and mean July temperatures were no higher than those in November at present. Understanding such events and their effects on the landscape can make a day in the Scottish mountains a thrilling excursion that spans millions of years.

This chapter provides some basic concepts and terminology for readers with limited prior knowledge of geology and geomorphology. It introduces the concept of recycling of rock materials, focusing on the ways in which rocks are broken down and eroded, and how rock material (sediment) is transported by various agencies

**Figure 1.1** Examples of the diversity of Scotland's mountain landscapes. (**a**) Suilven (731 m) in NW Scotland, a sandstone mountain rising above glacially scoured gneiss (photograph by John Gordon). (**b**) Sgùrr Dubh Mór (944 m) and Sgùrr Dubh an Da Bheinn (938 m), Cuillin Hills, Skye. (**c**) The eastern Grampians, with the Cairngorms on the skyline. (**d**) The Southern Uplands: White Coomb (821 m) from Hart Fell.

and deposited. It also outlines the geological timescale and describes how rocks, landforms and sediments can be dated and interrogated to reconstruct histories of past events.

## Geology and geomorphology

Geology is the science of the Earth and its history, reconstructed from the record of rocks: their types, structures, ages and origins. For over two centuries, geologists have studied the characteristics and distributions of rocks to piece together a coherent interpretation of how the present landmass of Scotland has evolved. Many of the pioneers of modern geology were Scots or focused their research in Scotland. It was here that James Hutton (1726–1797) formulated a governing principle of geology, now known as *uniformitarianism*: that the rock record is the product of the same processes as those that

now operate at the surface and subsurface of the Earth, or as Archibald Geikie (1835–1924) succinctly put it, 'the present is the key to the past'. In his influential treatise *Principles of Geology* of 1833 another Scot, Charles Lyell (1797–1875) consolidated the uniformitarian underpinnings of modern geology, providing the foundation for our present understanding of the geological record. Yet even these giants of nineteenth-century geology would have been astonished to learn that Scotland represents not a single landmass that has been present throughout geological time, but the agglomeration of fragments of ancient continents that have split asunder and collided as the tectonic plates of the Earth moved across the surface of the globe. They would have been equally astonished to find that over the past billion years the component parts of Scotland travelled from the Equator to near the South Pole, then slowly migrated northwards to their present position.

Many readers may be less familiar with the term *geomorphology*, which is derived from three Greek words meaning 'the science of the shape of the Earth'. Geomorphology is the study of landforms and the processes responsible for their formation. Essentially (but not exclusively), geomorphology focuses on the processes operating at the surface and near-surface of the Earth to produce landforms ranging in scale from millimetres (such as glacially abraded scratches on rock surfaces known as *striae*) to hundreds of kilometres (such as uplift and erosion of mountain ranges). From the evidence provided by landforms and the sediments deposited by rivers, glaciers, landslides and wind, geomorphologists attempt to reconstruct environmental conditions during the comparatively recent geological past, much as geologists use the rock record to reconstruct much older events in the Earth's history. Striae on rock surfaces, for example, tell us not only that a glacier was formerly present and the direction of glacier movement, but also that the base of the former glacier was at melting point and capable of sliding over and abrading the underlying rock surface (Fig. 1.2).

The poet William Blake (1757–1827) wrote that 'to see the world in a grain of sand' is 'to hold infinity in the palm of your hand'. If we could interrogate a sand grain, it might tell of being deposited in an ancient sea, cemented with other grains into solid rock, recrystallized by heat and pressure, then uplifted within a mountain chain. Eons may pass before the mountains are eroded and our sand grain forms part of a boulder deposited by a landslide; the climate cools, and the boulder is entrained then dumped by glacier ice, split by frost, and its fragments removed by a flood, broken down to sand by abrasion on a riverbed, until ultimately the grain of sand is again deposited in the sea. Locked in our grain of sand is, if not infinity, then at least a history that stretches back hundreds of millions of years. The story of our sand grain is cyclic: it starts and finishes on the sea floor. This continuous recycling of Earth materials is termed the *rock cycle* or *geological cycle*, a concept that owes its origins to the insights of James Hutton, who stared into the abyss of geological time and could find 'no vestige of beginning, no prospect of an end' in the constant recycling of Earth materials.

**Figure 1.2** Striae on the south summit (998 m) of Ben Challuim, near Tyndrum. These show that the last Scottish Ice Sheet flowed southeastwards across the mountain.

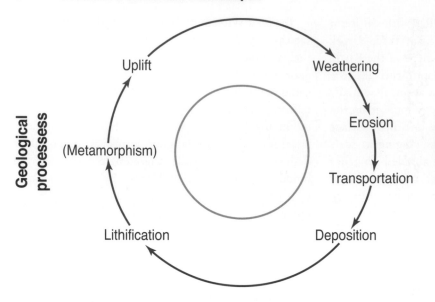

**Figure 1.3** The geological cycle. Sediments undergo lithification to form rocks that may be modified by recrystallization (metamorphism) and uplifted to form mountains. Ultimately even the highest mountains are reduced by the geomorphological processes of weathering, erosion, transportation and deposition of sediment.

As you can see from Figure 1.3 this cycle involves both geological and geomorphological processes. It begins with rock formation (lithification) through compaction and cementation of grains. Then follows, though not inevitably, changes induced in the rock by great heat or pressure (metamorphism), and ultimately in uplift to form alpine-scale mountain chains. Thereafter the geomorphological processes take over: weathering, which involves the breakdown of rocks by mechanical or chemical processes; erosion, the entrainment and removal of rock material by landslides, rivers, glaciers and wind; transportation of rock material by the same agents, and finally deposition of sediment. However, as the history of our sand grain illustrates, erosion, transportation and deposition may occur many times, by a variety of processes and routeways, until rock material reaches its final destination. Our hypothetical landslide boulder, for example, may have awaited the coming of glacier ice for hundreds of thousands of years, during which it was progressively reduced by weathering; the boulder may then have been transported and redeposited by several successive glaciers or ice sheets over a million years or so; once eventually entrained in a river during an exceptional flood it may have been redeposited on a gravel bar to await onward transport by the next great flood. So what geomorphologists call 'source-to-sink' transport of sediment is rarely an uninterrupted journey from the mountain to the sea: there are long intervals of stasis when the sediment is said to be 'stored', for example as landslide runout debris, in glacially deposited moraines or river floodplain deposits. These sediment stores represent distinctive landforms from which we can deduce the most recent formative history of a mountain landscape.

## Weathering, erosion and deposition
### Weathering

The term *weathering* is used to describe the breakdown of rock at or near the ground surface. *Physical weathering* is the disintegration of rock as a result of mechanical stress and *chemical weathering* is the progressive decomposition of minerals within a rock by chemical reactions. Both physical and chemical weathering processes tend to operate in combination: fracturing of rock by physical weathering enhances the access of water, the essential ingredient of all chemical weathering processes, and chemical weathering of minerals weakens rocks so that they are more likely to fracture or disintegrate.

Various physical weathering processes have affected the rocks that underlie Scottish mountains, but two are of particular importance: *stress release* and *frost weathering*. Most rocks originate deep underground under the weight of overlying rock. When the overlying rock is removed by erosion, a process that requires millions of years, the release of overburden weight can cause near-surface rock to fracture. More important in a Scottish context is the stress release that occurs comparatively rapidly (over a few millennia) as an ice sheet thins and disappears. Relieved of the weight of the ice, rocks tend to expand slightly, and such expansion results in cracking of rock to form *joints* or fractures. The formation of *stress-release*

**Figure 1.4** Bouldery sandstone debris near the summit of Maol Chean-dearg (933 m), Torridon. This debris formed through frost weathering prising open joints in the underlying bedrock, releasing boulders. These have been rounded by granular disaggregation, caused by ice forming in rock pores and detaching grains of rock. The white boulder in the foreground is a quartzite erratic that was dumped on the debris cover by the last ice sheet about 16,000 years ago.

*joints* is particularly important in weakening steep rock slopes, ultimately resulting in rockfalls and landslides.

Frost weathering operates in two ways. First, because the volume of ice is 9% greater than that of water, freezing of water in joints applies pressure to fracture walls, prising apart the rock. Similarly, water freezing in rock pores detaches grains from exposed rock surfaces, a process known as *granular disaggregation* (Fig. 1.4). Secondly, water freezing within porous rocks can form lenses of ice within the rock and growth of such ice lenses can split intact rock. Frost weathering therefore operates at two scales, producing coarse debris in the form of angular *clasts* (the term used to describe detached rocks of any size from gravel to large boulders) but also rounding the exposed edges of such clasts through granular disaggregation.

Chemical weathering is more insidious. The primary minerals that make up rocks, such as quartz, mica and feldspar, are chemically stable and largely insoluble. However, slow chemical reactions can alter insoluble minerals to more soluble products that are removed by groundwater, a process termed *leaching*. The rainfall over Scottish mountains is usually slightly acidic, but the acidity of water percolating through soil and peat is increased by incorporation of organic acids produced by decay of plant matter. This slightly acidic water slowly reacts with some minerals, gradually removing soluble products. The insoluble residue of such gradual reactions

often takes the form of secondary *clay minerals* that are chemically inert but mechanically weak, so chemically weathered rocks have limited resistance to physical weathering or erosion.

### Erosion

*Erosion* is a collective term for a wide range of geomorphological processes, all of which involve detachment or entrainment of rock, debris or soil. All forms of landslide involve erosion of slopes, as does soil erosion by *surface wash*, the movement of water over unvegetated soils. *Fluvial erosion* occurs when floodwaters entrain sediment from a river channel or riverbanks; at moderate streamflow velocities small particles (sand, silt and clay) are eroded, but at greater discharges gravel and even boulders may become entrained by rivers, bouncing and rolling along the channel floor until the flow velocity diminishes. Similarly, gusty mountain winds can entrain silt- to sand-sized particles and blow these downwind. As described in more detail in Chapter 4, *glacial erosion* occurs by abrasion of bedrock and detachment of rock fragments that become frozen onto the base of glacier ice and carried away.

### Deposition

*Deposition* is also a collective term, used to describe the dumping, emplacement or immobilization of material carried by landslides, rockfalls, surface wash, rivers, glaciers or wind. The nature of the material deposited by these various agencies differs, however, and plays a major role in allowing geomorphologists to reconstruct the effect of past events on the landscape. Catastrophic landslides deposit large angular boulders, some the size of cars; rockfall debris comprises boulders that accumulate at the foot of cliffs as *talus* (scree); surface wash deposits layers of sand and fine gravel; rivers deposit *stratified* (layered) sediments of silt to gravel size; glaciers deposit *till* (Chapter 4), which consists of stones and boulders embedded in a matrix of finer material; and as wind velocity drops, windborne *aeolian deposits* accumulate on the ground surface, often trapped by vegetation cover.

Because each of these deposits carries a genetic signature, sediments visible in a *section* (a vertical exposure, such as an eroding riverbank or roadcut) can usually be readily identified. Forensic examination of such sections can often reveal much more about the event that led to sediment deposition. This involves two related approaches: *sedimentology* (study of the characteristics of the deposited material) and *stratigraphy* (study of the relationships between a sediment unit and overlying or underlying sediments).

Though you will be familiar with the word 'sediment' (as, for example, the residue that accumulates in a bottle of venerable wine), geomorphologists regard all deposited materials as sediments, even large boulders. Sediments fall into two categories, grains and clasts. *Grains* are particles smaller than 2 mm. This category includes clay (particles smaller than 0.002 mm), silt (0.002–0.06 mm) and sand (0.06–2.0 mm). *Clasts* are generally classified as gravel (2–60 mm), cobbles (60–200 mm), boulders (200–600 mm) and large boulders (>600 mm).

Weighty textbooks have been written about sedimentology and stratigraphy, but a few examples will suffice to illustrate how these sediment properties aid interpretation of deposits and past events. One property that helps define the origin of a particular sediment is the degree to which it is *sorted* into grains or clasts of similar size. Windblown sand on a beach is well sorted, as all grains are of similar size. Gravel on a river bar is usually moderately sorted, exhibiting various clast sizes but within a fairly narrow range. Most glacial deposits, however, are extremely poorly sorted, as they contain particles ranging in size from clay to boulders. Another important criterion for differentiating deposits of different origin is *stratification* or layering. This occurs in fluvial sediments, where sediments of different size are deposited sequentially, so that, for example, a layer or *bed* of gravel deposited during a flood may be overlain by sand deposited as the flow velocity decreases. Windblown sands sometimes exhibit thin bedding, but those lacking bedding are termed *massive*, implying a lack of stratification. Glacial deposits are usually massive, though some types exhibit crude stratification.

Analysis of stratigraphy allows us to reconstruct a sequence of events. Figure 1.5 depicts a section where lake-floor sediment is overlain by till. This sequence tells

**Figure 1.5** An example of stratigraphy in Strath Bran, near Achnasheen. The lower, dark brown unit consists of silt deposited on the floor of a glacier-dammed lake that covered Achnasheen about 12,500 years ago. The upper yellowish unit is a glacial till that was deposited as the glacier advanced into the lake.

us that a glacier crossed the valley, damming up a lake, then advanced over the lake deposits. Similar principles apply to ancient sediments that are now lithified to form solid rock. In the Midland Valley of Scotland, for example, there are repeated threefold sequences of *sedimentary rocks*, those comprising sediments originally deposited in water or on the land surface, then subsequently lithified. At the base of each sequence are limestones, which are overlain in turn by sandstones then by coal seams. Study of the fossils in the limestones indicates that these were derived from corals and other marine fauna that lived in a warm, shallow sea. The bedding in the overlying sandstones tells us of gradual infilling of this ancient sea by sandy deltas deposited by rivers. The coal deposits represent subsequent emergence of a swampy low-lying plain colonized by dense forests of primitive trees and other plants that were later buried by sediment, converted to peat then altered to coal under immense heat and pressure. Thus the industrial revolution in Scotland was powered by ancient forests of long-extinct trees, the fossilized remains of which are today preserved in Fossil Grove, in Victoria Park in Glasgow.

## Dating the past: the geological timescale

The Earth is about 4.54 billion years old, a timespan that has been subdivided into a hierarchy of units of varying duration. Because the geological timescale is so long, we use millions of years as the basic unit of geological time, and thousands of years to describe the age of more recent events during the last two million years. 'Years' are denoted by *a* (Latin *annus*), thousands of years as *ka* and millions of years as *Ma*; these abbreviations may refer to a span of time but are mainly employed in the sense of 'time before present'.

The initial subdivision of geological time was made by nineteenth-century geologists, who established a *relative* timescale of geological periods based on the fossils that occur in rocks of different ages, the boundaries between different geological periods being denoted by extinction of some diagnostic species and the emergence of new species. Our current use of geological periods (Cambrian, Ordovician, Silurian etc; Fig. 1.6) to subdivide geological time developed in this way but had the drawback that it could be applied only to rocks

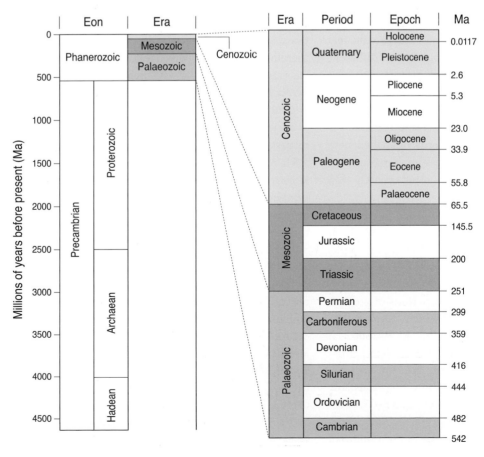

**Figure 1.6** The geological timescale. Note that the timescale on the right is compressed for the Palaeozoic and Mesozoic and expanded for the Quaternary.

bearing the fossils of animals with hard shells. These first appeared during the Cambrian period (542–482 Ma), so all older rocks were initially lumped into a category of 'Precambrian' age, which represents about 88% of the time since formation of the Earth. About a century ago, however, geologists began to experiment with using the radioactive decay of particular elements in rock-forming minerals to establish an *absolute age* for rocks, an approach referred to as *radiometric dating*.

The principle underlying radiometric dating is that some elements in rock-forming minerals exist in an unstable form (an unstable *isotope*) that very slowly decays to form a different element: the original element is known as the parent, and the decay product is known as the daughter. The rate of decay is relatively rapid at first, then slows progressively through time, following a mathematical rule called a negative exponential function, and the time taken for an isotope to decay to half of its original mass is known as its *half life*. For example, the mineral zircon occurs as crystals in some rocks, and contains the parent unstable isotope uranium-235 ($^{235}U$), which progressively decays to form the daughter isotope lead-207 ($^{207}Pb$). The half life of $^{235}U$ is 704 Ma, so if half of the original mass of $^{235}U$ is present in a zircon crystal, the age of rock is 704 Ma; if a quarter of the original mass of $^{235}U$ is present, the age of the rock is approximately ($2 \times 704$ Ma = 1408 Ma) and so on. Several different parent–daughter combinations with different half lives are employed in radiometric dating of rocks, providing independent checks of the results. All rock ages obtained by radiometric dating contain uncertainties, which is why the timings of geological events described here are often prefaced by the symbol '~', implying 'approximately': thus '~425 Ma' implies 'approximately 425 million years ago'.

Radiometric dating of rocks has allowed not only the provision of absolute ages to define the age and duration of the classical geological periods (thus, for example, Cambrian 542–482 Ma; Ordovician 482–444 Ma), but also dating of much older rocks and subdivision of the Precambrian timescale. The longest periods of geological time are referred to as *eons*, of which there are four (Hadean, ~4540–4000 Ma; Archaean, ~4000–2500 Ma; Proterozoic, ~2500–542 Ma; and Phanerozoic, 542 Ma–present). These are subdivided into *eras*: for example, the Phanerozoic eon is subdivided into the Palaeozoic era (542–251 Ma), the Mesozoic era (251–65.5 Ma) and the Cenozoic era (65.5 Ma–present). Each era is further subdivided into periods, such as the Cambrian

and Ordovician periods, and periods are subdivided into *epochs*. Figure 1.6 shows the main components of this scheme. We live in the Holocene epoch (11.7 ka to the present) of the Quaternary period (the past 2.6 Ma), which is part of the Cenozoic era (the past 65.5 Ma), a subdivision of the Phanerozoic eon (the past 542 Ma), though some scientists argue that we have recently entered a new geological epoch, the Anthropocene, dominated by the environmental effects of human activity.

## Dating the Quaternary

The most recent geological period, the Quaternary, began about 2.6 Ma. As described in Chapter 4, the Quaternary represents the period known as *the* Ice Age, though the Quaternary Ice Age is only the most recent of several ice ages that have punctuated the long history of planet Earth. The Quaternary is subdivided into two geological epochs: the Pleistocene (2.6 Ma to 11.7 ka) and the Holocene (11.7 ka to the present). Most landforms in the Scottish mountains are of comparatively recent (late Quaternary) origin. The main reason for this is that the last (most recent) Scottish Ice Sheet, which culminated around 27–22 ka, obliterated almost all earlier mountain landforms apart from major erosional features that had progressively evolved during successive glaciations. Most of the landforms described in chapters 5–10 therefore relate to the period following retreat of the last ice sheet, and particularly the period after ~17 ka when Scottish mountains began to emerge from the thinning and retreating ice.

As with rocks, stratigraphy is useful in establishing the relative age of Quaternary deposits or landforms (Fig. 1.5). A similar approach applies to landforms and is known as *morphostratigraphy*. If landslide runout debris has buried a moraine, for example, the landslide must have occurred after retreat of the glacier that produced the moraine.

Three approaches have been used to establish the absolute age of events, landforms and deposits within the period since ~27 ka, when the last ice sheet began to retreat. The first is *radiocarbon dating*. Living organisms contain the same concentration of the unstable isotope carbon-14 ($^{14}C$) as the atmosphere. When a plant or animal dies, the concentration of $^{14}C$ in the dead tissue declines by 1% every 83 years. As a result, only 50% of the original $^{14}C$ is present after 5730 years (the half life of $^{14}C$), 25% after 11460 years, 12.5% after 17190 years and so on. By measuring the concentration of $^{14}C$ in dead organic matter, we can therefore establish the time of death or burial. This technique has been useful for establishing, for example, the age of buried peat or other organic material, or when vegetation started to colonize terrain following glacier retreat. However, because the concentration of $^{14}C$ becomes very small with increasing age, radiocarbon dating is usually accurate only for ages up to about 45 ka, or exceptionally 60 ka. Moreover, because atmospheric $^{14}C$ varies slightly through time, radiocarbon dates differ slightly from 'true' ages, and require to be calibrated to a calendrical timescale. All radiocarbon ages cited in this book are calibrated ages.

Secondly, of increasing importance in a Scottish context is *terrestrial cosmogenic nuclide exposure dating*, here abbreviated to *TCN dating* but sometimes referred to as *cosmogenic isotope dating*. When a rock or boulder surface becomes exposed at the ground surface as a result of a landslide, or through retreat of a glacier or thinning of an ice sheet, it becomes subject to bombardment by cosmic radiation. This causes distinctive 'cosmogenic' nuclides (isotopes) to progressively accumulate within some minerals, so by measuring the concentration of such nuclides in the surface layers of a rock outcrop or boulder we can determine how long it has been exposed. The main nuclides used to determine exposure age are beryllium-10 ($^{10}Be$) and chlorine-36 ($^{36}Cl$); the former is employed on quartz-rich rocks, and the latter used on quartz-poor rocks such as basalt. However, because the rate of accumulation of cosmogenic nuclides in rocks is very slow (just four atoms per gram per year for $^{10}Be$), this method suffers from relatively poor precision, though dating precision and confidence can be improved by dating several samples from the same site.

Finally, *optically stimulated luminescence dating* (*OSL dating*) works on the principle that buried sediments are subject to very low levels of natural radiation that cause electrons to become trapped within the crystal lattices of mineral grains. Shining a beam of light onto a sample under controlled laboratory conditions causes an emission of light that is proportional to the concentration of trapped electrons and thus the age of sediment deposition.

All of these dating techniques experience some drawbacks and all give only an approximate 'true' age. In scientific papers the precision or uncertainty of an individual age is conventionally shown as '±' after the 'best estimate' age. For example, an age cited as 15.7±0.2 ka implies that there is a 68% probability that the true age occurs between 15.9 ka and 15.5 ka, and a 95% probability that the 'true' age occurs between 16.1 ka and 15.3 ka.

In general, radiocarbon dating has much greater precision than TCN or OSL dating. To avoid such complications, ages in this book are simply shown as approximate: thus an age of 15.7±0.2 ka is shown as ~15.7 ka, implying 'approximately' 15,700 years ago.

## About this book

Most books and articles devoted to the evolution of the Scottish landscape are targeted at the professional research community of geologists and geomorphologists, and assume familiarity with the scientific terms and concepts of their disciplines. In particular, papers published in scientific journals tend to be dense, highly technical, littered with citations, punctuated by equations and expressed in the arcane argot of Earth science. At the other extreme are popular accounts, typically shorn of jargon and replete with imaginative colour reconstructions of past landscapes (often featuring dinosaurs, volcanoes or glaciers) but limited scientific content.

This book attempts to steer a middle course. Its aim is not only to describe the landforms and landscapes of Scottish mountains, but also to provide the background for understanding and appreciating the evolution of Scotland's mountain scenery, and to give some idea of how our present understanding has been achieved. Because some readers may have limited knowledge of the bewildering terminology employed in the dark arts of geology and geomorphology, each chapter starts by outlining the concepts and terms necessary for understanding the topics that follow.

It is appropriate to conclude this introduction with a caveat. A popular notion is that science advances when discovery builds on discovery as part of an undeviating journey towards complete understanding. This is nonsense. Scientists interpret the evidence available to them in what appears to be the most logical fashion, but the appearance of new evidence or theoretical perspectives may leave their conclusions in ruins. This is particularly true of geology and geomorphology, where the record of past events is often fragmentary and the discovery of a new exposure showing inverted strata, the appearance of a particular fossil in the 'wrong' rocks, or a suite of aberrant radiocarbon ages can overturn years of painstaking research and lead to the search for different interpretations. At the end of the twentieth century, for example, most geomorphologists were persuaded by the evidence then available that the last Scottish Ice Sheet was of rather limited extent and that some higher mountain summits remained above the level of the ice. But cracks soon appeared in this interpretation in the form of conflicting evidence, and in the present century this interpretation collapsed completely under assault from new technologies that demonstrated that the ice sheet buried all Scottish mountain summits and extended far out across the continental shelves that surround Scotland. In less than a decade, our view of the size of the ice sheet roughly doubled in terms of both its extent and its thickness.

As this example demonstrates, our understanding of Scotland's geology and geomorphology is subject to constant revision. If you are reading this book 20 years after its publication some of the interpretations outlined here will certainly have been modified or superseded. Moreover, there are a number of current controversies concerning aspects of the evolution of Scotland's mountain landscapes, so I have had to make some interpretative choices that others will challenge. Such challenges are to be welcomed, for they stimulate our attempts to understand not just the mountain scenery of Scotland, but the evolution of planet Earth itself.

# Chapter 2

# The geological evolution of Scotland

## Introduction

To non-geologists, geological maps might appear to be a bewildering series of brightly coloured blotches; take away the outlines and they resemble masterpieces of abstract art. Interpretation is not helped by the arcane terminology employed by geologists to describe rocks. Consider Figure 2.1: if you know what is implied by 'Neoproterozoic Moine metasedimentary rocks' then you are either a geologist or have sufficient geological understanding to choose geology as your specialist subject on Mastermind. We can demystify this terminology slightly by noting that each unit in the key contains two descriptors: first the age of the rock, denoted by the geological period it belongs to (such as Jurassic or Neoproterozoic; Fig. 1.6) and then the type of rock, such as granite or sandstone.

For a small country 80,240 km$^2$ in area, Scotland contains a staggering range of rock types of widely different ages, the result of a complex and sometimes violent geological history. To understand this geological diversity, it is useful to first identify the major pieces in the jigsaw puzzle of Scotland's geology. These are termed *terranes*, zones of the Earth's crust that preserve a geological history that differs from adjacent areas. Scotland comprises five terranes: the Hebridean terrane, Northern Highlands terrane, Grampian Highlands terrane, Midland Valley terrane and Southern Uplands terrane. As can be seen from Figure 2.2, these are separated by major *faults* that represent vertical and horizontal movement of one terrane relative to its neighbour: thus the Great Glen Fault separates the Northern Highlands terrane from the Grampian Highlands terrane, and the Highland Boundary Fault separates the Grampian Highlands terrane from the Midland Valley terrane. If you compare Figures 2.1 and 2.2, you will see that each terrane contains rocks of different ages or types from those in other terranes, though some rocks such as granites occur in several terranes. Also cutting across terrane boundaries are the relatively young (61–55 Ma)

rocks of the Hebridean Igneous Province, which represent the result of intense volcanic activity in western Scotland, the most recent episode of rock formation now widely represented in the landscape. Before we examine the evolution of these terranes, we need to consider first the nature of the main rock types, then the principles of plate tectonics, which have been responsible for creating and joining the five Scottish terranes.

## Rocks
### Minerals

Rocks are aggregates of *minerals*, naturally occurring crystalline solids with a regular internal structure. Some rocks consist of just one type of mineral, but most comprise several mineral species. Granite, for example, contains the common minerals quartz, feldspar and mica. Of the thousands of different minerals that exist naturally, the most common are the *silicates*. Quartz, one of the most common and stable silicate minerals, consists of silicon dioxide ($SiO_2$) or silica. Feldspars, the commonest minerals in the Earth's crust, comprise silicon, oxygen and potassium or calcium plus sodium. Various other common minerals such as mica, olivine and pyroxene are also members of the silicate group. Another important group comprises the carbonate minerals, such as calcite (calcium carbonate, $CaCO_3$), the main constituent of chalk and limestone, and dolomite (calcium-magnesium carbonate). Geologists recognize three major categories of rock, differentiated by their origins: igneous rocks, sedimentary rocks and metamorphic rocks.

### Igneous rocks

An *igneous rock* is one that has cooled and crystallized from molten material derived from beneath or within the Earth's crust. We can distinguish two types: *extrusive* (or *volcanic*) rocks, formed where *magma* (a mixture of molten rock and gas) has erupted and flowed as lava onto

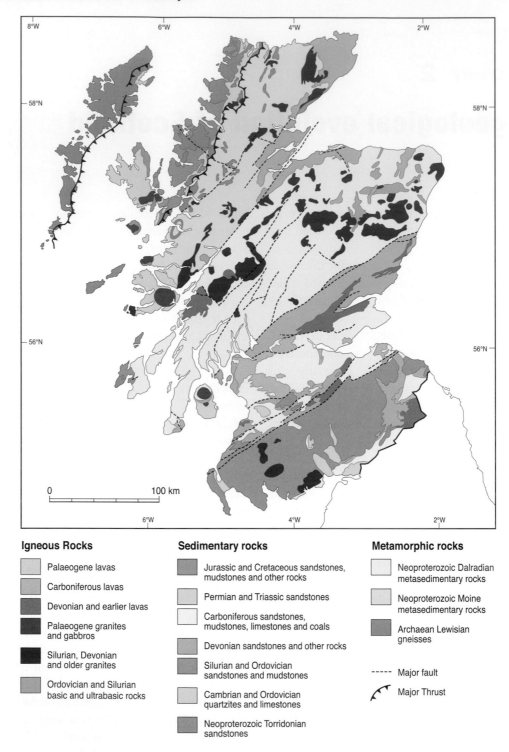

**Igneous Rocks**

- Palaeogene lavas
- Carboniferous lavas
- Devonian and earlier lavas
- Palaeogene granites and gabbros
- Silurian, Devonian and older granites
- Ordovician and Silurian basic and ultrabasic rocks

**Sedimentary rocks**

- Jurassic and Cretaceous sandstones, mudstones and other rocks
- Permian and Triassic sandstones
- Carboniferous sandstones, mudstones, limestones and coals
- Devonian sandstones and other rocks
- Silurian and Ordovician sandstones and mudstones
- Cambrian and Ordovician quartzites and limestones
- Neoproterozoic Torridonian sandstones

**Metamorphic rocks**

- Neoproterozoic Dalradian metasedimentary rocks
- Neoproterozoic Moine metasedimentary rocks
- Archaean Lewisian gneisses

- - - - -  Major fault

Major Thrust

**Figure 2.1** Geological map of Scotland. Based on BGS geological data (permit No CP19/009) © UKRI 2019. All rights reserved.

the surface, where it has cooled and solidified rapidly; and *intrusive* (or *plutonic*) rocks, formed where molten rock has been injected under a thick cover of overlying rocks, so that it has cooled and crystallized very slowly, often in the form of a huge mass of subterranean rock called a *pluton* (Fig. 2.3). The red blobs in Figure 2.1 represent granitic plutons that have been exposed at the surface (*unroofed* in geological jargon) by erosion of the

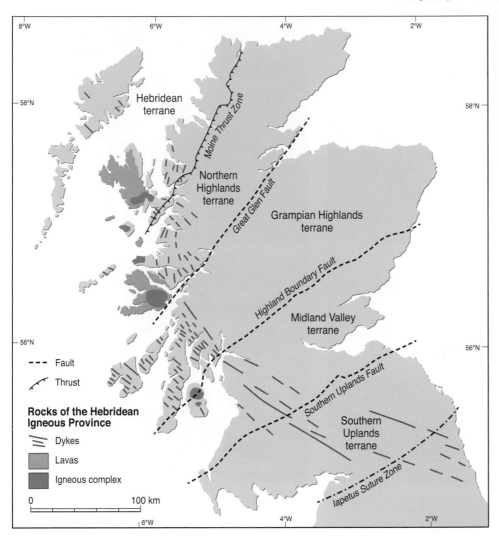

**Figure 2.2** The five terranes of Scotland, the major faults that separate these terranes, and the terrestrial distribution of Palaeocene igneous complexes, lava fields and dykes that comprise the Hebridean Igneous Province.

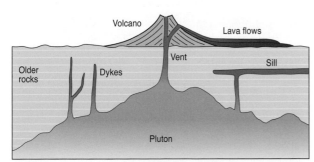

**Figure 2.3** Igneous rock formations. The volcano and lava flows are composed of extrusive (volcanic) rocks, whereas the pluton, dykes, sill and vent comprise intrusive (plutonic) rocks.

overlying rocks. Rapid cooling of molten rock, as in lava flows, produces very small mineral crystals, but very slow cooling of deep subterranean molten rock allows the formation of much larger crystals. This distinction is important for rock climbers, as coarse crystalline (plutonic)

rocks such as granite and gabbro provide much better grip (frictional contact) than lavas such as basalt (Fig. 2.4).

Where magma has been injected underground along the bedding planes of sedimentary rocks it forms a *sill*, a sheet-like *intrusion* that interrupts the sedimentary sequence (Fig. 2.3). Sill rocks are usually more resistant than the underlying and overlying sedimentary rocks, and consequently form steep scarps when exposed by erosion. The Lomond Hills scarp in Fife, for example, is the exposed margin of a thick sill, and Salisbury Crags in Edinburgh represent another. Where magma has been intruded into vertical fissures it solidifies to form *dykes*. In locations where dyke rocks are more resistant than the adjacent rocks, they form wall-like structures, typically 1–10 m wide, such as those along the south coast of Arran. Where dykes have proved less resistant to erosion

**Figure 2.4** Some Scottish rocks. (**a**) Granite outcrop on Arran. (**b**) Two eroded basalt dykes form trenches in gabbro, Sgùrr Alasdair, Skye. (**c**) Cliff of stacked basalt lava flows, Trotternish, northern Skye. (**d**) The contact (unconformity) between grey pebbly basal Cambrian quartzite and the underlying pink Torridonian sandstones. (**e**) Moine schists. Erected vertically by folding, the bands open along foliation planes like the leaves of a book. (**f**) Boulder of banded Lewisian gneiss resting on Torridonian sandstone boulders.

than the surrounding rocks, as in the Cuillin Hills, they form vertical trenches or chimneys (Fig. 2.4b). Amongst climbers and hillwalkers, however, the most infamous

effect of dyke erosion has been isolation of the narrow spire of fine-grained gabbro that forms the Inaccessible Pinnacle (Fig. 2.5), the highest point (986 m) on Sgùrr

**Figure 2.5** The Inaccessible Pinnacle, a spike of fine-grained gabbro that forms the highest point (986 m) of Sgùrr Dearg on the Cuillin ridge, Isle of Skye.

Dearg and the nemesis of many 'Munro baggers' who seek to climb all Scottish summits over 3000 feet (914 m). Figure 2.2 depicts in a generalized way the *dyke swarm* that developed during Palaeogene igneous activity (~61–55 Ma).

Igneous rocks are classified by their mineral composition. Whereas granites consist mainly of quartz, feldspar and mica, gabbros are largely composed of olivine, pyroxene and calcium-rich feldspar. Similarly, rhyolite lavas are rich in quartz and feldspar, and represent the fine-grained equivalent of granite, whereas basalt lavas have a mineral composition similar to that of gabbro; lavas of intermediate composition are called andesites. There is, however, a spectrum of mineral composition in igneous rocks, from those that are relatively silica-rich, and referred to as *acidic* or *felsic* rocks (such as granite and rhyolite) to those that are intermediate (such as andesite, and diorite or syenite, its plutonic cousins) to those that are relatively silica-poor and referred to as *basic* or *mafic* rocks (such as gabbro, basalt and dolerite). Much rarer than these in Scotland are *ultrabasic* (or *ultramafic*) rocks composed almost entirely of olivine and pyroxene.

## Sedimentary rocks

Most sedimentary rocks are *clastic sedimentary rocks*, composed of particles derived from the weathering and erosion of pre-existing rocks. Such particles are transported to their site of deposition by water (rivers, waves and tidal currents), glaciers or wind. As they accumulate they are gradually transformed into solid rock (*lithification*) by compaction under the weight of overlying sediment and cementation by a secondary mineral, such as silica or calcite, that precipitates out of percolating groundwater. Clastic rocks are classified on the basis of the size of their constituent particles. The coarsest, consisting of gravel to boulder size fragments, are conglomerates, often originally deposited by high-energy rivers. Sandstones are rocks composed mainly of quartz grains where the dominant particle size is that of sand (0.06–2.0 mm). Variants include *arkose sandstones* (or *arkoses*), coarse sandstones containing abundant feldspar fragments, and *greywackes*, muddy sandstones originally deposited on the ocean floor by submarine slumps and currents. The Torridonian sandstones of northwest Scotland are arkoses (Fig. 2.4d), and most of the rocks in the Southern Uplands are weakly altered greywackes. Most sandstones were originally deposited by rivers, either on land, in alluvial fans or floodplains, or on the floor of lakes or the sea; others are of aeolian (wind-deposited) origin, and originally formed dunes in arid environments. Fine-grained sedimentary rocks containing grains finer than sand include siltstones, mudstones and shales, which are differentiated by their mineral content; most fine-grained rocks were originally deposited as sediment on lake floors or the seabed. Clastic sedimentary rocks usually form beds or *strata* that are separated from overlying and underlying layers by *bedding planes*. The layered appearance of mountains such as Liathach and Suilven (Fig. 1.1) in northern Scotland represents Torridonian sandstone strata separated by bedding planes.

A separate category of sedimentary rocks consists of *biochemical sediments*, such as some limestones and coal. Organic limestones are composed of the calcite-rich 'shells' or skeletons of molluscs or corals that have accumulated on the seabed. Limestones also form by chemical precipitation of calcite from seawater, or through accumulation of fragments of pre-existing limestones, and many limestones have a composite (organic, chemical and/or detrital) origin.

## Metamorphic rocks

Rocks classified as *metamorphic* are those that have recrystallized in a solid state by the effects of heat and pressure on pre-existing rocks deep within the Earth's crust, though if the temperature is high enough and water-rich fluids are present some components (feldspars and quartz) may partially melt. The pre-existing rocks may be of igneous, sedimentary or metamorphic origin. The terms *metasediments* or *metasedimentary rocks* are used to describe sedimentary rocks that have undergone metamorphism: most of the rocks forming the mountains of the Scottish Highlands are of this type. Metamorphism operates at two scales. *Regional metamorphism* involves widespread alteration and recrystallization of rocks during the deformation that accompanies the growth of mountain chains, such as the Caledonian Mountains that formed the precursor of much of the Highlands. *Contact metamorphism* is more localized and occurs where pre-existing rocks come into contact with magma, as around the margins of a granite pluton.

If a rock is composed of a single mineral, the metamorphosed equivalent is independent of the heat and pressure involved. Thus, for example, sandstone comprising silica-cemented quartz grains is recrystallized into quartzite, and limestone is metamorphosed into marble. The quartzite that underlies summits in northwest Scotland, such as those of Beinn Eighe and Arkle, is a metasedimentary rock with recrystallized silica cement. Where several minerals are present in the original rock, however, there is a *metamorphic grade sequence* that reflects progressive increases in temperature and pressure. For example, low-grade metamorphism alters shales or mudstones to slate, medium-grade metamorphism produces phyllite, medium- to high-grade metamorphism produces schist, and high-grade metamorphism alters the original rock to gneiss. Schists underlie much of the Northern Highlands and Grampian Highlands terranes and are commonly subdivided into pelitic schists (or *metapelites*) that represent metamorphosed shales or mudrocks, and psammitic schists (or *metapsammites*) that represent metamorphosed sandstones. Schists are typically *foliated* (minerals have a common parallel orientation), rich in mica and banded (Fig. 2.4e) though because the rocks have been folded and metamorphosed the banding does not always represent the original bedding. Gneisses also occur in these terranes and form the surface rocks across much of the Hebridean terrane; these are characteristically coarse-grained and banded, exhibiting alternation of *schistose* (mica-rich) bands and *granulite* bands dominated by quartz and feldspar crystals (Fig. 2.4f). In contrast, the greywackes of the Southern Uplands terrane have been only slightly altered through low-grade metamorphism to form a tough rock in which the finer-grained constituents still preserve fossils.

The metamorphic rocks of the Northern Highlands terrane belong to the *Moine Supergroup* and comprise metamorphosed sediments that were originally deposited in a shallow sea 980–875 million years ago; these are mainly metapsammites and metapelites derived from sandstones with occasional mud and shale layers. Those of the Grampian Highlands terrane belong mainly to the *Dalradian Supergroup* and are largely derived from marine sediments deposited 730–480 million years ago. These are more varied, being derived from a wider range of sedimentary rocks, and comprise mainly schists, quartzite, gneiss, slate and marble.

## Stratigraphy

As outlined in Chapter 1, *stratigraphy* is the study of the relative age of rocks, particularly sedimentary rocks, based on their relationship to each other and their fossil content. Unless sedimentary rocks have been overturned by later folding or older rocks have been pushed over younger rocks by *thrust faulting*, in any depositional sequence the strata 'young' upwards: those at the base of a sequence are oldest and those at the top are the youngest. A second tenet of stratigraphy is that sediments are initially deposited horizontally, for example on the sea floor, so that tilted sedimentary rocks represent later folding. This principle however, is vitiated by some sediments, such as those deposited on the steep frontal slope of a delta. A third stratigraphic principle is that fragments of pre-existing rocks within sediments must be derived from older rocks that pre-date the sediments. Finally, the principle of cross-cutting relationships holds that strata that cut across a younging-upwards sequence of rocks must be younger than the rocks they cut across. Such cross-cutting often represents an *unconformity*, or time gap in the geological record, where older rocks have been folded and eroded prior to the deposition of younger sediments (Fig. 2.4d). The fossil content of a stratigraphic sequence not only provides an approximate age for the rocks but may also exhibit changes that reflect changing environments of deposition, or an evolutionary sequence. Similarity in fossil assemblages also allows sedimentary rocks in one location to be correlated with those in another.

## Plate tectonics

The origin of the five Scottish terranes and the Hebrides Igneous Province is explicable by *plate tectonics*, a concept that underpins our understanding of global geological evolution. In this context, the term *tectonics* refers to large-scale vertical or horizontal movements of the Earth's crust, such as the migration of continents across

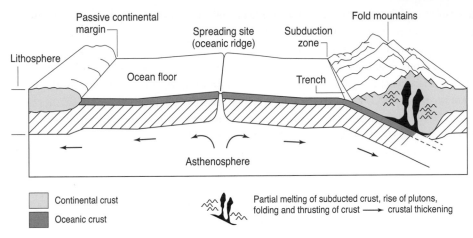

**Figure 2.6** Near-surface structure of the Earth, showing the lithosphere (crust plus rigid upper mantle) and the underlying ductile asthenosphere. Also shown are a divergent plate boundary (spreading site) in mid-ocean, a passive continental margin and a continental-margin convergent plate boundary (subduction site). Landward of the subduction site, folding of rocks and intrusion of plutons is causing crustal thickening and mountain uplift.

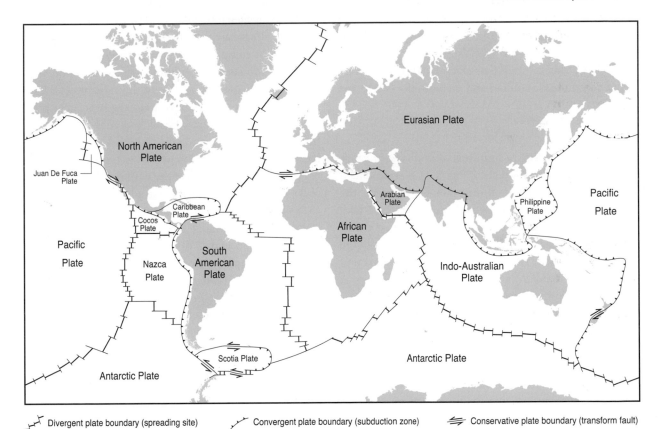

**Figure 2.7** The fourteen major tectonic plates, showing the locations of spreading sites, subduction zones and conservative plate boundaries (transform faults).

the globe or the formation of mountain chains. *Tectonic plates* are large, rigid segments of the *lithosphere*, which incorporates the Earth's crust and the uppermost part of the underlying *mantle*, a zone that typically extends to depths of 50–100 km but more under large mountain ranges. Underlying the lithosphere is the *asthenosphere*, a zone of ductile (deformable) rock in a partially molten

state (Fig. 2.6). The rigid tectonic plates are therefore 'floating' on the asthenosphere, and thus capable of moving very gradually across the surface of the Earth. At present the lithosphere is divided into 14 major tectonic plates of different sizes (Fig. 2.7).

The Earth's crust has two distinct components: *continental crust*, which has an average thickness of about

37 km, but is up to 85 km thick under geologically recent mountain chains such as the Himalayas; and *oceanic crust*, which is typically 5–7 km thick. Continental crust is composed of a wide variety of igneous, metamorphic and sedimentary rocks, but is dominantly acidic (felsic) in composition, with an average density of ~2.7–2.9 g cm$^{-3}$ and contains rocks up to ~4000 Ma in age. Oceanic crust is composed of basalt, with an average density of ~3.0 g cm$^{-3}$ and is nowhere older than ~200 Ma. Because oceanic crust is denser than continental crust, the lithosphere underlying oceans 'floats' at a lower level in the underlying asthenosphere than that under continents (Fig. 2.6). This is why the oceanic crust is submerged under the sea, typically to a depth of 3–5 km, whereas most continental crust rises above present sea level. The main exceptions are *continental shelves*, which are the low-lying margins of continents, submerged under up to 200 m of water. The British Isles are surrounded by a broad continental shelf that extends 80–100 km west of the coastline of Scotland and includes the North Sea Basin.

## Plate margins

Considered over tens to hundreds of millions of years, the pattern of tectonic plates (Fig. 2.7) is transient. If we go back about 200 million years, for example, the Atlantic Ocean did not exist and the Americas, Africa and much of Europe formed a single landmass called *Pangaea* that has subsequently fragmented into the present continents. Fast forward 200 million years and it is likely that the Atlantic Ocean will be shrinking, ultimately reuniting the continents at its margins. The process driving the birth, growth, decline and demise of oceans is convection in the outer part of the Earth's mantle. Heat is conducted through the mantle from the Earth's core, and in the outer part of the mantle (the asthenosphere) forms convection cells, much like those that form in a saucepan containing thick soup that is heated from below. The overlying lithosphere is rafted on these vast cells of slowly moving ductile rock, causing continents to split apart (*continental rupture*), migrate gradually across the Earth's surface (*continental drift*) and ultimately to collide and bond (*continental suture*). The manner in which this occurs can be seen by investigating the nature of plate boundaries.

*Divergent plate boundaries* are those where two plates are moving apart above the upward-moving and separating limbs of upper-mantle convection cells. This process may begin under a continent, causing heating, thinning and uplift of the crust. As the two slabs of crust slowly move apart, the intervening terrain subsides to form a *rift valley*, such as the East African Rift Valley today. Further divergence of the two slabs is caused by injection and eruption of molten basaltic magma from the upper mantle, which tends to push the two slabs farther apart until a narrow 'proto-ocean' is formed, such as the Red Sea between Africa and Arabia. If injection and eruption of magma continue, the ocean gradually widens through addition of new basaltic oceanic crust along the site of the original rift, driving the two parts of the ruptured continent farther apart. Ultimately a mature ocean such as the present Atlantic Ocean develops, with a divergent plate margin or *spreading site* along its central axis, where basaltic magma is both erupted from submarine volcanoes and injected into dykes, adding to the oceanic crust (Fig. 2.6). Because the lithosphere at such sites is hot and therefore of comparatively low density, spreading sites form vast submarine mountain ridges, such as the Mid-Atlantic Ridge, with a rift along the ridge crest where new rock is being added and spreading is occurring. As the oceanic lithosphere spreads and cools, however, it becomes denser and gradually sinks to form deep ocean basins, typically up to 5 km below sea level. This process of *ocean floor spreading* has been confirmed by dating the rocks of the oceanic crust: these are geologically young (a few million years) near the spreading site but get progressively older towards the ocean margins.

The addition of new oceanic crust at divergent plate boundaries implies that there is compensatory loss of crust elsewhere. This occurs at *convergent plate boundaries* where oceanic lithosphere is returned to the mantle. All oceans contain the seeds of their own destruction: as the oceanic lithosphere gradually spreads and cools, it ultimately reaches the density of the underlying asthenosphere and begins to sink, forming a deep oceanic trench near the ocean margin. This return of oceanic lithosphere into the mantle is termed *subduction*, and can be traced by the depths of earthquakes, which show that the descending plate initially dips at about 45° under the over-riding plate, becoming steeper with depth. Some subduction sites occur at continental margins (Fig. 2.6), such as present-day subduction of the Nazca Plate under the western margin of the South American Plate (Fig. 2.7); others occur at *island arcs* near ocean margins, such as the Japanese island arc. The onset of subduction creates a new convergent plate boundary and explains why oceanic crust never exceeds ~200 Ma in age. As the rate of subduction at ocean margins eventually exceeds the rate of spreading at oceanic ridges, all oceans are doomed to slowly shrink, bringing the continents at their margins closer together, and ultimately die, as the oceanic lithosphere separating continents is completely consumed back into the mantle and the continents bordering the oceans collide.

Finally, *conservative plate boundaries* are those where two plates meet but crust is neither created or destroyed. Such boundaries typically take the form of a zone of *transform faults*, where one plate is sliding past an adjacent plate. Such movement is not continuous but takes the form of 'stick-slip' sudden lateral movement that generates major earthquakes. The San Andreas fault zone in California and the Alpine Fault in New Zealand, for example, mark transform plate boundaries. Both may be relatively dormant for decades or centuries, before rupturing and triggering potentially devastating earthquakes.

## Mountain building

The fundamental principle underpinning our understanding of mountain formation is that of *isostasy*. Imagine blocks of wood floating in water (Fig. 2.8a). Low density blocks float high above the water surface, but the proportion of each block that is submerged remains constant no matter how thick the block is. Denser blocks sit lower in the water, because each block displaces an equivalent mass of water; the blocks are in *isostatic equilibrium* with the water. As the lithosphere (crust plus upper mantle) 'floats' on the ductile asthenosphere, it behaves in roughly the same way. If the continental crust undergoes thickening, as described below, it adjusts isostatically by both developing a deeper 'submerged' root and by rising at the surface to form mountains until it regains isostatic equilibrium (Fig. 2.8b). An interesting corollary of isostasy is that if an existing mountain range is eroded and thereby loses mass, it will experience renewed uplift, though to a lower altitude than before. This means that erosion of mountains is a very prolonged process because they continue to rise isostatically even as they are being eroded. It also means that the deep roots of mountain ranges ultimately constitute the high ground, after all overlying rock has been removed. Much of the Scottish Highlands comprises metamorphic rocks and granite plutons that formed deep underground but have now been uplifted to form the mountains of the Northern Highlands and Grampian Highlands terranes.

The formation of major mountain chains such as the Alps or Andes is termed *orogenesis* and occurs at

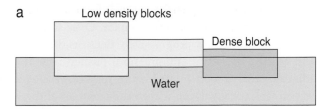

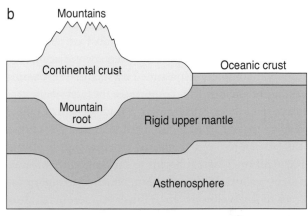

**Figure 2.8** The principle of isostasy. (**a**) Wooden blocks floating in water. The denser block sits lower in the water than the less dense blocks. (**b**) Lithosphere (crust plus rigid upper mantle) in isostatic equilibrium, 'floating' above the ductile asthenosphere. The denser oceanic crust sits lower than the continental crust, and a deep mountain root supports the overlying mountains.

convergent plate margins due to three processes, all of which cause crustal thickening and isostatic uplift. The first is *crustal shortening* due to lateral compression of rock at such boundaries. This compression causes the rocks to rumple up to form folds, causing a marked increase in rock volume per unit area; this is why major mountain belts at convergent plate margins are often referred to as *fold mountains*. Compression at convergent boundaries also causes *thrust faulting* of huge bodies of rock so that they are pushed up on top of each other, again thickening the crust. The second mechanism involves partial melting of the descending plate or base of the over-riding plate at convergent boundaries, so that relatively low-density magma, often granitic in composition, rises up into the crust (Fig. 2.6). Much of this magma gradually solidifies underground to form plutons, causing crustal thickening, but some is erupted to form a *volcanic arc*, a chain of large volcanoes (*stratovolcanoes*) composed mainly of rhyolitic or andesitic lavas interbedded with ash layers. The best-known volcanoes on Earth, such as Vesuvius and Mount St Helens, are stratovolcanoes, and notorious for their violent eruptions. Thirdly, as an oceanic plate is subducted, ocean-floor sediments are scraped off its surface and accumulate at the margin of the over-riding plate to form an *accretionary complex*. Over time the accretionary complex progressively thickens, is folded and faulted by lateral compression, pushed over the adjacent crust and often intruded by magma, causing it to rise isostatically. Deep within the crust, all of these processes are accompanied by metamorphism, mainly of sedimentary rocks, forming rocks such as schist and quartzite.

Mountain building reaches its most extreme form when the oceanic lithosphere between two adjacent continents is completely consumed by subduction, causing *continental collision*. This compresses the intervening sediments and rocks, causing intense folding, thrust faulting of huge masses of rock and marked crustal thickening. The base of the over-thickened crust reaches depths of 100–150 km, where it begins to melt and form magma, usually granitic in composition, which rises within the crust to form plutons. The Himalayan mountain chain represents an example of such collision orogenesis, formed by the meeting of India and southern Asia as the intervening ocean was completely subducted.

Most of the mountains in the Scottish Highlands and Southern Uplands owe their origins to ancient episodes of mountain building at converging and colliding plate margins but have subsequently been eroded so that the roots of the original mountains have emerged at the surface. Such mountain-building periods are referred to as *orogenies*. The mountains of southern Europe such as the Alps, for example, were uplifted during the geologically recent Alpine Orogeny, caused by the collision between Africa and Europe. As outlined below, the ancestral mountains of much of the Highlands and Southern Uplands were formed during a much older and rather complex mountain-building event, the Caledonian Orogeny, which spanned the period ~490–390 Ma.

## Geological evolution: the making of Scotland

The study of geological evolution is sometimes compared with that of historians trying to reconstruct past events from fragmentary documentary evidence. In one sense, however, this analogy is flawed: historians can assume that the physical geography of the land they study (the distribution of land, sea, coastlines, rivers and mountains) is fairly constant over their comparatively brief timescales of investigation. By contrast, geologists not only have to reconstruct past events from the fragmentary rock record, but also have to place these in the context of

the shifting geography of planet Earth caused by the rupture, migration and collision of tectonic plates. Here the geological evolution of Scotland is introduced first by considering a broad perspective of Scotland's changing position relative to former continental landmasses, then by detailing the main events affecting each of Scotland's five terranes.

## The big picture: shifting continents and vanished oceans

Reconstructing the former configuration of continents and oceans becomes increasingly challenging as we go back in geological time, in part because the rock record becomes more fragmentary, but also because the oldest rocks are mainly gneisses that have undergone repeated metamorphism, making their evolutionary interpretation difficult. On the Outer Hebrides are some of the oldest rocks on Earth, gneisses that formed 3125–3000 million years ago, but detailed understanding of Scotland's location on the changing configuration of continental landmasses is largely limited to the past 600 million years or so.

Figure 2.9 depicts snapshots of Scotland's location relative to continental landmasses for eight periods. Our story starts around 760 Ma, when the vast supercontinent of Rodinia began to break up. Further rifting of this huge landmass around 600–580 Ma separated the ancient continents of Laurentia (North America and Greenland) and Baltica (Scandinavia) from the supercontinent of Gondwana, which incorporated present-day South America, Africa and Antarctica (Fig. 2.9a). By ~550 Ma an ancient ocean, the Iapetus Ocean, separated these three landmasses, with what are now the northern parts of Scotland located near the margin of Laurentia (Fig. 2.9b). By ~510 Ma, the Iapetus Ocean had reached its maximum width of over 2000 km and began to shrink as island arc subduction sites developed around its margins, causing gradual reconvergence of Laurentia, Baltica and Gondwana. The island arc that developed near the margin of Laurentia, known as the Midland Valley Arc (Fig. 2.9c) was drawn towards Laurentia by subduction and collided with the continent around 470 Ma. By the same time, the ancient continent of Avalonia, which supported England and Wales, had split from Gondwana and was carried towards Laurentia as the Iapetus Ocean contracted. The stage was set for the tectonic convulsions that created much of the geology of Scotland.

The period from ~490 Ma to ~390 Ma was critical in the formation of the present-day landmass of Scotland

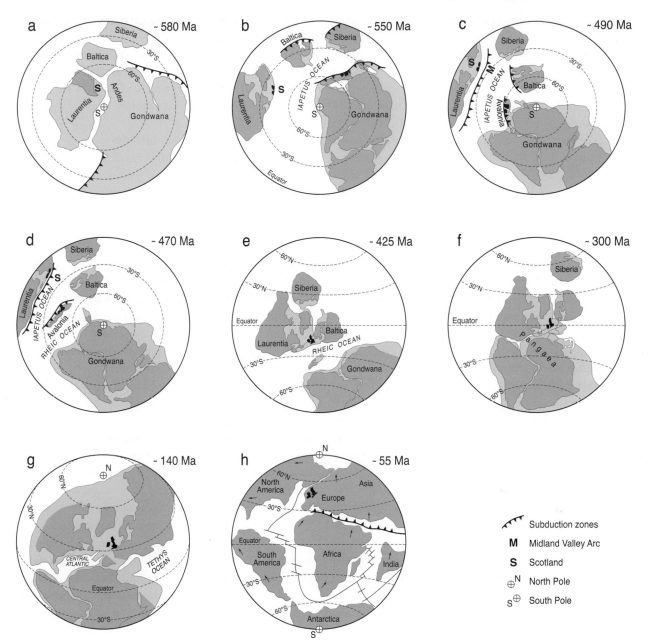

**Figure 2.9** Snapshots of Scotland's position relative to continental landmasses at eight times in the past, from ~580 Ma to ~55 Ma. During this long period Scotland migrated from near the South Pole to about 50°N, crossing the Equator at ~300 Ma. The five terranes of Scotland were not united until ~425–400 Ma. Partly adapted from Torsvik, T.H. *et al.* (1996) Continental break-up and collision in the Neoproterozoic and Palaeozoic – a tale of Baltica and Laurentia. *Earth-Science Reviews* 40, 229–258. © 1996 Elsevier with permission from Elsevier.

as this was when the five terranes depicted in Figure 2.2 slid laterally into approximately their present positions along northeast–southwest trending fault lines. During this long period the shrinkage and eventual demise of the Iapetus Ocean pulled Laurentia, Baltica and Avalonia together – and Scotland was caught between all three. The resulting Caledonian Orogeny occurred in two stages. The earlier, known as the Grampian orogenic

event, occurred around 480–465 Ma, by which time the Midland Valley Arc had collided with Laurentia (Fig. 2.9d), and subduction under what is now central and southern Scotland had initiated the formation of fold mountains in the Grampians and Northern Highlands. Around 450 Ma Baltica collided obliquely with Laurentia, sliding laterally in the process and initiating the Scandian orogenic event, which culminated at ~435–425 Ma. In northwest Scotland the immense pressure generated by this collision pushed the entire Caledonian mountain belt westwards over the much older rocks of the Hebridean terrane along a complex of thrust faults to form the Moine Thrust Zone, which now marks the boundary between the Hebridean and Northern Highlands terranes. (Fig. 2.2). Also during the Scandian event, the oblique convergence of Laurentia and Avalonia was marked by the folding and thrusting of the thick accretionary complex of oceanic sediments that now form the Southern Uplands terrane. The final disappearance of the Iapetus Ocean was represented by a comparatively 'soft' collision between Laurentia and Avalonia that resulted in the geological union of Scotland with England along the Iapetus Suture zone, which runs from the Solway Firth to just south of the Cheviot Hills, roughly parallel to the present political boundary between the two countries (Fig. 2.2).

By ~425 Ma, the now-united parts of Great Britain were located at about latitude 15°S, inland from the southern margin of the reunited continents of Laurentia and Baltica (a landmass variously referred to as 'Euramerica' or 'Laurussia') and separated from Gondwana by the Rheic Ocean (Fig. 2.9e). Subduction and closure of the Rheic Ocean reunited 'Laurussia' with Gondwana, creating the supercontinent of Pangaea by ~300 Ma (Fig. 2.9f). Closure of the Rheic Ocean was accompanied by another major orogeny, the Hercynian Orogeny of ~380–280 Ma. In Europe this resulted in the formation of fold mountains in a belt stretching from northern Spain to Turkey, but its effects on Scotland were limited to comparatively gentle faulting and folding of rocks, particularly in the Midland Valley.

The final chapter in the tectonic history of Scotland was underway by ~140 Ma, at which time rifting and ocean-floor spreading had begun to split Pangaea into southern and northern components, separated by the proto-Atlantic Ocean in the west and Tethys Ocean in the east (Fig. 2.9g), and the North Sea Basin was opening between Scotland and Scandinavia. Initially, Scotland was little affected by these events, but when rifting extended into what is now the North Atlantic region, Scotland experienced its last rock-forming event as volcanoes erupted along its western seaboard, spewing out copious volumes of basaltic lava to form the Hebridean Igneous Province (61–55 Ma). As the North Atlantic Ocean widened, the locus of igneous activity shifted gradually from our shores, and Scotland settled into relative tectonic quiescence, drifting slowly northwards and

eastwards on a passive continental margin (Fig. 2.9h) as part of the Eurasian plate. The story, however, does not end with the sputtering extinction of the last Hebridean volcano, as Scotland had yet to experience episodes of plateau uplift and tilting during the Cenozoic era, events that are outlined in the next chapter.

## Evolution of the Hebridean terrane

Of the five terranes in Scotland (Fig. 2.2), the Hebridean terrane is unique as it was largely unaffected by the Caledonian Orogeny. It contains three very distinct groups of rocks: Lewisian gneisses, Torridonian sandstones and Cambro-Ordovician sedimentary rocks.

The Lewisian gneisses are high-grade metamorphic rocks that underlie almost all of the Outer Hebrides and a zone along the northwestern seaboard from Cape Wrath to Knoydart, as well as the islands of Coll, Tiree and Iona, southeast Skye and southwest Islay (Fig. 2.1). They include the oldest rocks in Europe, with some dating back to at least ~3125 Ma on the Outer Hebrides. The Lewisian gneisses are divided into a northern region dominated by pink quartz-feldspar-mica gneiss intruded by numerous granitic veins, and southern and central regions underlain by grey banded gneiss and granulite with layers and pods of basic and ultrabasic igneous rocks (Fig. 2.4f). These three regions are thought to have originally formed separate fragments of the ancient continent of Laurentia that were brought together by ~1700 Ma. Most of the rocks comprising the Lewisian gneisses were originally igneous in origin and formed before ~2700 Ma but subsequently experienced folding, metamorphism, partial melting and intrusion by dykes and granitic rocks deep within the crust during at least two long, multi-phase episodes of intense deformation and metamorphism.

The Lewisian rocks are often termed the *basement* rocks of Scotland as they form the stable platform over and against which other terranes developed as a result of continental collision during the Caledonian Orogeny. This basement demonstrably underlies parts of the Northern Highlands terrane, where Lewisian-like ('Lewisianoid') gneisses have locally been exposed by erosion of the overlying younger (Moine) rocks but is also present at depth in the more southerly terranes.

The Lewisian gneisses are believed to have been uplifted to the surface between 1400 Ma and 1200 Ma, and formed a platform of low ground and mountains on which the Torridonian sandstones accumulated. These occur in a discontinuous belt along the northwest seaboard from Cape Wrath to Rum (Fig. 2.1), are the oldest unaltered sedimentary rocks in Britain and are represented by three groups. The oldest Torridonian rocks form the Stoer Group and Sleat Group, both of which represent sedimentary piles deposited directly on the Lewisian basement. The Stoer Group comprises a ~2 km thickness of arkosic sandstones, conglomerates and siltstones that accumulated prior to ~1200 Ma in rift valleys. The rocks of the Sleat Group occur in the southern part of the Torridonian outcrop and consist of a ~3.5 km thick stack of greyish sandstones and mudstones that were originally deposited by rivers in shallow lakes or seas. The Sleat Group is unconformably overlain by the rocks of the Torridon Group. These began to accumulate around 1000–970 Ma, eventually achieving a thickness of 5–6 km by ~800 Ma, consist of arkosic (feldspar-rich) red sandstones with occasional conglomerate beds, and were originally deposited on alluvial fans and plains. The rocks of this group form most of the great Torridonian sandstone mountains of northwest Scotland, such as Suilven, Slioch, Ben Alligin and Liathach.

There followed a long (~270 Ma) interval, during which the Torridonian rocks were extensively eroded and gently tilted before being drowned under the shallow tropical seas of the Laurentian continental shelf on the margin of the Iapetus Ocean (Fig. 2.9b). Around 530 Ma, quartz sand began to accumulate on the shelf, covering Lewisian gneisses and Torridonian sandstones alike (Fig. 2.4d). This now forms the youngest rock on the mountains of northwest Scotland, a distinct white or grey band of Cambrian quartzite that underlies several summits, such as those of Foinaven, Arkle and Beinn Eighe (Fig. 2.10). On low ground the quartzites are overlain by Cambro-Ordovician shales, grits and limestones, but these are absent from high ground.

The Hebridean terrane is bounded to the east by the Moine Thrust Zone (Fig. 2.2), along which the rocks of the Moine Supergroup were pushed over the rocks of the Hebridean terrane during the final episode of the Caledonian Orogeny. The structure of the thrust zone is locally complex, and in some places successive slices of westward-thrust rocks are superimposed and folded. As some of these slices incorporate bodies of Lewisian, Torridonian and Cambro-Ordovician rocks, the orderly sequence of rocks is locally dislocated: rafts of Lewisian gneiss have been carried over Torridonian rocks, 'thrust' Torridonian rocks overlie Cambrian quartzites, or there are multiple sequences of the same rock, as near Ben More Assynt, where successive rafts of Cambro-Ordovician rocks are stacked above one another.

**Figure 2.10** Cambrian quartzite (grey) overlying Torridonian sandstones south of Glen Torridon. The unconformity separating the two represents a time gap of over 270 million years.

## The Northern Highlands terrane

The rocks of the Northern Highlands terrane between the Moine Thrust Zone and the Great Glen Fault are dominated by schists of the Moine Supergroup; these are locally intruded by granitic rocks and overlain in Caithness and along the coast of the Moray Firth by younger sedimentary rocks, mainly siltstones and sandstones of Devonian age (Fig. 2.1). The schists were originally deposited as a thick sequence of sandy, silty and muddy sediments in a shallow, subsiding marine setting on the margin of Laurentia within the period 980–870 Ma. The timing of sediment deposition therefore overlaps with that of the Torridonian sandstones, but the Torridonian and Moine rocks appear to have accumulated in widely separated areas that were subsequently pushed together along the Moine Thrust Zone.

The Moine schists include both metapsammites and metapelites, and are typically banded (foliated) as a result of the growth of parallel layers of mica, a phenomenon known as *schistosity* (Fig. 2.4e). Structurally, the metamorphic rocks of the Moine Supergroup represent complex thrust slices that have been pushed together and over each other, and as a result the schists are strongly folded and locally interleaved with 'Lewisianoid' gneisses. The original marine sediments were initially deformed and metamorphosed during tectonic events centred around 800 Ma and 740 Ma before being intruded by granitic rocks around 600 Ma. The final chapter in the evolution of the Moine Supergroup was written during the Caledonian Orogeny. The effects of the Grampian orogenic event of ~480–465 Ma are difficult to decipher, as they have been overprinted by those of the later Scandian event of ~435–420 Ma. During this later event, collision between Baltica and Laurentia resulted in renewed metamorphism, folding and thrusting of Moine rocks, accompanied by further intrusion of plutons, crustal thickening and uplift. It was at this time that Moine rocks were pushed westward by at least 70 km (and probably more) across the rocks of the Hebridean terrane to form the Moine Thrust Zone. This remarkable event involved westward movement of thick slices of rock deep below the surface along an array of interconnected thrust planes. Subsequent erosion of several kilometres of overlying rock has now revealed these thrusts, notably at Knockan Crag near Elphin, where the Moine Thrust separates Moine rocks from the underlying Cambro-Ordovician rocks.

## The Grampian Highlands terrane

The Grampian Highlands terrane is dominated by the metamorphic and igneous rocks of the Dalradian Supergroup. These comprise a huge thickness of repeatedly folded and deformed rocks, including slates, phyllites, schists, quartzites, limestones and metamorphosed igneous rocks that underlie most of this region apart from areas around the Moray Firth, where they are overlain by younger sedimentary rocks (Fig. 2.1).

The history of the Dalradian rocks began with rifting and separation of Laurentia from Gondwana, leading to the opening of the Iapetus Ocean (Figs 2.9a and 2.9b). This was accompanied by deposition of marine sediments in subsiding offshore basins, forming the Dalradian succession of sandstones, siltstones, mudstones and limestones over the period ~730–480 Ma; the Dalradian succession is therefore much younger (and more varied) than that of the sedimentary rocks that ultimately formed the Moine Supergroup. Rifting and opening of the Iapetus Ocean was also accompanied by igneous activity around 600 Ma, now evident in lavas of that age in Argyll and the intrusion of granite plutons into the thickening sediment pile.

During the Grampian orogenic event (~480–465 Ma), the sedimentary and igneous rocks of the Dalradian succession experienced extensive metamorphism and repeated deformation episodes that folded and re-folded the rocks, causing crustal thickening and uplift of the Caledonian Mountains. The axes of the earliest folds ran approximately east–west, but later compression produced complex overfolds (*recumbent folds* or *nappes*) aligned northeast to southwest, so that in some areas the rocks are vertical or completely overturned. The Dalradian rocks that underlie Ben Lawers, for example, are upside down, having been completely folded over; the overlying 'rightside up' parts of the nappe have been completely removed by erosion. Some granites, such as the Aberdeen granite in northeast Scotland, were intruded at this time. The main phase of igneous activity in the Grampians, however, occurred soon after the Scandian orogenic event (~435–425 Ma), when crustal thickening resulting from the collision of Laurentia and Baltica caused melting at the base of the crust and rise of molten plutons that now form most of the granite masses of the Grampian Highlands (Fig. 2.1). These are referred to as the 'newer' granites and were emplaced between 425 Ma and 400 Ma, after deformation of the Grampian metasediments was complete. Such 'newer' granites underlie some of the highest mountains in the Grampian Highlands, such as the Cairngorms, Lochnagar and those around upper Loch Etive.

Of similar age are the volcanic rocks of Lorne and Lochaber in the western Grampians. These are part of a large volcanic complex that erupted on numerous occasions to create a pile of mainly basaltic lavas up to 800 m thick, though these locally show an upwards transition to andesitic and rhyolitic lavas such as those of the Ben Nevis and Glen Coe volcanic complexes.

### The Southern Uplands terrane

The Southern Uplands terrane is bounded in the north by the Southern Uplands Fault and in the south by the Iapetus Suture Zone (Fig. 2.2), the latter marking the zone where England was grafted on to Scotland as the Iapetus Ocean was finally consumed by subduction, bringing Avalonia into contact with Laurentia. Most of the rocks of the Southern Uplands terrane originated as muddy sandstones (*turbidites*) deposited by successive large slumps across submarine fans on the floor of the Iapetus Ocean during the Ordovician and Silurian periods (~470–420 Ma). Each slump deposited first a thick layer of muddy sand, overlain by thinner layers of silt then clay as fine particles settled. This pattern was repeated many times, forming successive beds in which the particles fine upwards. Lithification of the muddy sandstones produced greywackes: coarse-grained, poorly-sorted sandstones, often grey, black or purplish in colour. These typically contain angular grains of quartz, feldspar, dark silicate minerals and fragments of igneous or metamorphic rocks, all embedded in a muddy matrix. Recrystallization of this matrix by low-grade metamorphism has produced a tough, resistant rock that is interbedded with siltstones, mudstones or shales derived from the original silt and clay layers.

Structurally, the Southern Uplands represent an accretionary complex that accumulated at the margin of Laurentia. As the oceanic crust of the shrinking Iapetus Ocean was subducted under the Laurentian plate, the thick sediments that had accumulated on the ocean floor were scraped up and pushed northwards across the Laurentian margin. This process generated massive lateral compression, and as a result the sediments were folded, thrust along faults, and the greywackes became weakly metamorphosed. The Southern Uplands terrane is subdivided into three zones, separated from each other by northeast–southwest trending faults: a northern belt (~470–460 Ma), which contains volcanic rocks

derived from collision of an island arc with Laurentia; a central belt (~460–430 Ma) and a southern belt (~430–420 Ma). The formation of the Southern Uplands terrane therefore spans most of the Caledonian Orogeny.

During the final stages of the collision between Laurentia and Avalonia at ~400–390 Ma, melting of the base of the thickened crust resulted in intrusion of granites, now represented by the Loch Doon, Carsphairn, Criffel and Cairnsmore of Fleet granites of southwest Scotland, and by the Cheviot granite and volcanic complex, the latter being geologically associated with the Southern Uplands terrane even though the granite outcrop lies just across the border in England. These granite intrusions represent the closing events of the Caledonian Orogeny in Scotland.

### The Caledonian Mountains and the Old Red Sandstones

At this point it is useful to summarize the sequence of events that occurred during the Caledonian Orogeny, place this in wider context and picture what Scotland looked like when the orogeny terminated around 390 Ma. We have seen that the earliest (Grampian) events occurred around 480–465 Ma, with subduction at the margin of Laurentia and collision of the Midland Valley Arc with the Laurentian plate. The second major phase, the Scandian event of ~435–425 Ma, resulted from oblique collision of Baltica with Laurentia. The closure of the Iapetus Ocean finally grafted England on to Scotland as Avalonia collided with Laurentia.

At the beginning of the orogeny the segments of Laurentian continental crust that now underlie the terranes of Scotland were geographically separated, probably by several hundreds of kilometres. During the orogeny, and particularly during the final closure of the Iapetus Ocean, they slid gradually together along the major faults that delimit terrane boundaries (Fig. 2.11), reaching approximately their present positions relative to each other by ~420–400 Ma. It is also notable that though the Caledonian sedimentary and metasedimentary rocks of the Northern Highlands, Grampian Highlands and Southern Uplands formed over different timescales and in different tectonic settings, all were originally derived from thick accumulations of marine sediments, and all experienced folding and thrust faulting that caused crustal thickening, resulting in the uplift of mountains with deep crustal roots. In all three terranes the intrusion of plutons added to the crustal mass, promoting further isostatic uplift.

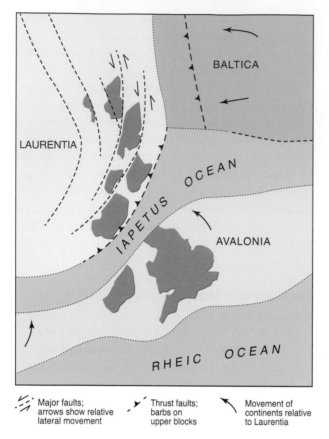

Major faults; arrows show relative lateral movement

Thrust faults; barbs on upper blocks

Movement of continents relative to Laurentia

**Figure 2.11** Reconstruction of the closure of the Iapetus Ocean and assembly of the Scottish terranes along major faults during the early Devonian (~400 Ma). Adapted from Soper, N.J. *et al.* (1992) *Journal of the Geological Society* 149, pp. 871–880 and Stephenson, D. *et al.* (1999) Caledonian Igneous Rocks of Great Britain: an introduction. *Geological Conservation Review Series* 17, pp. 1–26. Joint Nature Conservation Committee, Peterborough. © JNCC.

The Caledonian mountain chain was not confined to Scotland, but extended northwards along the boundary between Laurentia and Baltica, forming the mountains of Scandinavia and east Greenland, and westward along the boundary between Laurentia and Avalonia to form the Appalachian mountain chain in North America. Later plate movements have separated the components of this vast mountain range. We do not know what height the Caledonian Mountains achieved, but it is likely that they were at least as high as the European Alps (3000–5000 m) and some authors have suggested that they may have reached Himalayan proportions. At the time of their formation Scotland lay in an arid continental interior location towards the southern margin of the 'Laurussian' landmass 10–30° south of the Equator. As the mountains grew, they were subject to vigorous erosion. Violent earthquakes associated with crustal deformation must

have precipitated huge landslides, similar to those that have occurred during the geologically recent rise of the Southern Alps in New Zealand, and powerful rivers cut down into the mountains as they rose, transporting vast quantities of sediment that were deposited in adjacent low-lying basins that slowly subsided under the weight of the accumulating sediment.

These sediments now form the rocks of the Old Red Sandstone Supergroup and are identified as 'Devonian sandstones and other rocks' in Figure 2.1. The Old Red Sandstone rocks get their reddish hue from deposition in an arid environment, where rust-coloured haematite (a mineral form of iron oxide) formed the cement binding the sandstone grains. The main areas of outcrop occur in a broad belt south of the Highland Boundary Fault, and in Caithness, Orkney and around the Moray Firth. Less extensive outcrops occur in Shetland, the southern part of the Midland Valley and in southeast Scotland, where Old Red Sandstone strata overlie the greywackes of the Southern Uplands terrane. The original extent of these rocks, however, was much greater than their present outcrop suggests: residual fragments are preserved in both the Northern Highlands and Grampian Highlands terranes, and they underlie younger (Carboniferous) sedimentary rocks in much of the Midland Valley.

Deposition of these sediments began in the middle Silurian (~435 Ma), as the Caledonian mountains were uplifted, and continued into the Carboniferous period to ~350 Ma, though south of the Highland Boundary Fault sandstones of middle Devonian age (~400–370 Ma) are absent. Gradual subsidence of the Midland Valley resulted in sediments being deposited as large alluvial fans on which the sediments were coarsest adjacent to the former source areas and decreased in grain size away from former mountain margins. The Old Red Sandstones of Caithness, Orkney, and the Moray Firth area were mainly deposited during the middle and late Devonian (~400–360 Ma) in an extensive subsiding basin, the Orcadian Basin. Some of these are of terrestrial origin, deposited by rivers draining the Caledonian mountains. Others are of lacustrine origin, having been laid down in a large shallow lake of fluctuating extent.

The volume of sediment in the Old Red Sandstones and various other lines of evidence suggest that a thickness of 25–30 km of rock was eroded from the Caledonian mountains prior to ~400–380 Ma. This does not, of course, imply that the mountains reached 25 km in height. The apparent mismatch between the probable height of the mountains (3–5 km) and the volume of rock removed reflects two factors. First, erosion occurred as the mountains rose, so uplift was partly counteracted by rock removal. Second, as the Caledonian mountains were eroded, the reduction in mass caused isostatic uplift of the underlying crust, so that erosion of the rising mountains continually replenished the supply of sediment to adjacent areas. This second point is important, because it means that although the mountains formed during the Caledonian orogeny are the ancestors of those now occupying the Northern Highlands, Grampian Highlands and Southern Uplands, the rocks that now underlie the mountains in these areas represent the deep roots of the former Caledonian mountains. This is demonstrated by the fact that the granites of the Grampian Highlands and Southern Uplands were probably intruded at least 5–10 km below the Caledonian Mountains but by ~400 Ma had been exposed at the surface through complete removal of the overlying rocks. So rapid was the erosion of the Caledonian Mountains that by ~400–380 Ma the relief of the Highlands may have been no greater than that of today. The uplift and erosion of the Caledonian Mountains represents a classic example of the geological cycle (Fig. 1.3): the marine sediments of the Moine and Dalradian Supergroups experienced lithification, metamorphosis, and uplift to form part of the Caledonian mountain chain, then were weathered, eroded and ultimately deposited as the Old Red Sandstones.

The distribution of surviving Old Red Sandstone outcrops suggests that some of the major elements of the present landscape were already in place by ~400 Ma, notably the main Grampian watershed, major valleys draining towards the Moray Firth and a depression along the Great Glen Fault. These ancient topographic features have survived because major drainage routeways (valleys) have tended to persist throughout later periods of uplift, and also because subsequent periods of uplift appear to have favoured existing upland areas, preserving ancient topographic features that represent the ancestors of the present relief.

### Evolution of the Midland Valley terrane

The evolution of the Midland Valley terrane can be considered to have had two stages, the earlier occurring mainly during the Devonian period and the later during the Carboniferous period. Throughout this long interval Scotland moved northwards across the Equator and experienced a change from the predominantly hot arid conditions of the Devonian to the humid equatorial conditions of the Carboniferous.

As outlined above, during much of the Devonian period the Midland Valley subsided between the Highland Boundary Fault to the north and the Southern Uplands Fault to the south, and great thicknesses of Old Red Sandstone sediments were deposited in the subsiding basin. Around 415–410 Ma, the deposition of Old Red Sandstone sediments was accompanied by volcanic activity, and in some areas the sandstone strata are interdigitated with volcanic rocks. This activity represents a postscript to subduction of the Iapetus Ocean: a slab of subducted lithosphere broke off on its descent into the asthenosphere, causing partial melting at the base of the overlying crust and eruption of a first generation of Midland Valley volcanoes. These are now represented by thick accumulations of basaltic lava flows, although andesitic flows appear near the top of the lava pile. These lavas now form the Ochil Hills and Sidlaw Hills, and part of the Pentland Hills south of Edinburgh.

The evolution of the Midland Valley during the Carboniferous period was dominated by extensional tectonics, crustal stretching and thinning, and the formation of a complex rift valley between the Highland Boundary Fault and the Southern Uplands Fault. Shallow seas periodically flooded the subsiding rift but were infilled by sandy deltas that emerged to form lowlying forested swamps, a cycle that was repeated several times in various areas. The sediments deposited in these settings were lithified to form sequences of limestones, sandstones, siltstones, oil shales and coal deposits that were subsequently faulted and gently folded as a result of far-field effects of the Hercynian Orogeny.

Crustal stretching and thinning during the Carboniferous period was accompanied by a second generation of Midland Valley volcanic activity as magma produced near the top of the mantle erupted at the surface. The most extensive area of Carboniferous volcanic rocks occurs in an arc around Glasgow (Fig. 2.1), where successive near-horizontal basalt lava flows were erupted around 335 Ma and now form a stack of *plateau basalts* up to 1000 m thick that underlies the Fintry Hills, Campsie Fells, Kilpatrick Hills and Renfrewshire Heights (Fig. 2.12). Further evidence of Carboniferous eruptions takes the form of *volcanic plugs*, steep-sided hills that represent the resistant rocks that solidified in the vents of former volcanoes. These are dotted throughout much of the Midland Valley, and include such prominent landmarks as Dumbarton Rock, Dumgoyne on

**Figure 2.12** Stacked basalt lava flows of Carboniferous age on the southern scarp face of the Campsie Fells north of Glasgow.

the flanks of the Campsie Fells, the Lomond Hills in Fife, Edinburgh Castle Rock and North Berwick Law. These volcanic plugs represent only a very small part of the original volcanoes. Edinburgh Castle Rock and the two plugs that constitute nearby Arthur's Seat, for example, represent the skeletal remains of a volcano that was originally about 5 km in diameter and probably rose about 1000 m above the adjacent ground. Now all that remains is the resistant core of the volcanic edifice and the adjacent lava flows of Whinny Hill, the rest having been removed by erosion.

Igneous events in the Midland Valley persisted into the early Permian period (~290–280 Ma), marked mainly by the intrusion of dolerite sills into the Carboniferous sedimentary rocks (dolerite is a medium-grained igneous rock, compositionally similar to basalt). By far the largest of these is the Midland Valley Sill, emplaced around 307 Ma, which underlies an area of about 1900 km$^2$ around the inner Firth of Forth and crops out under Stirling Castle, forms the Lomond scarp in Fife and

supports the foundations of the Forth railway bridge, the heaviest steel engineering structure in Great Britain.

## The Hebridean Volcanic Province

Following the early Permian igneous activity described above, Scotland was undisturbed by cataclysmic events until the onset of the Palaeogene period at 65.5 Ma. During the Permian and Triassic periods (299–202 Ma) the British Isles moved slowly northwards from ~10°N to 30–35°N and experienced renewed hot arid conditions. Sedimentary rocks (mainly New Red Sandstones) that were deposited during this interval now have limited outcrop on land and nowhere form high ground, though they are extensive on offshore shelves. During the period ~210–70 Ma, Scotland experienced three long episodes during which rising sea levels engulfed much of the land surface, leading to widespread deposition of Jurassic and Cretaceous sedimentary rocks, but these have been largely removed as a result of subsequent erosion. Only patchy remnants occur on land, mainly on low ground around the Moray Firth and in the Inner Hebrides. The main tectonic event during this period was crustal stretching and rifting east of Scotland, which resulted in opening of the North Sea Basin. By the late Cretaceous, (~70–65.5 Ma) all of Scotland had probably been reduced to a landscape of low relief.

This long period of relative tectonic quiescence was abruptly terminated in the early Palaeogene, when a mantle plume (rising convection current) became established under what is now Iceland, causing uplift, crustal stretching, rifting, and eventually opening of the North Atlantic Ocean after ~55 Ma. In Scotland this event was marked by a comparatively brief (~61–55 Ma) episode of volcanic activity, now represented by igneous rocks on the islands of the Inner Hebrides as well as the Ardnamurchan Peninsula and southern Morvern. The most widespread consequence of this activity was the eruption of successive basalt lava flows, each flow being typically 5–15 m thick (Fig. 2.4c). Basalt lavas are particularly fluid, and these spread out over the underlying rocks almost horizontally, originally covering an estimated area of ~40,000 km² between the Outer Hebrides and Firth of Clyde; their much-eroded terrestrial remnants occur on and around Morvern, northern Skye, Mull, Rum, Canna, and Muck (Figs 2.1 and 2.2). Basaltic volcanicity was accompanied and succeeded by the intrusion of the central igneous complexes of Skye, Rum, Ardnamurchan, Mull and Arran during the interval 60–56 Ma. These represent the roots of ancient volcanoes and had differing individual histories, generally involving intrusion of magma and uplift of surrounding and overlying rocks (probably to heights of 2000 m or more), then partial collapse and unroofing by rapid erosion, which exposed their plutonic cores. Some igneous complexes are granitic in composition, such as those underlying the Red Hills on Skye and the mountains of northern Arran (Fig. 2.4a). Others consist of gabbros, such as those that underlie the Cuillin Hills on Skye (Fig. 2.4b).

Palaeogene volcanicity along the western seaboard was also accompanied by the development of a northwest–southeast trending dyke swarm (Fig. 2.2). Hundreds of dolerite dykes intruded older rocks in all of Scotland's terranes, and a few even reached northern England. The intrusion of the central igneous complexes may have had an even more long-lasting effect on the landscape, however, as it appears to have been accompanied by preferential uplift along the west coast, tilting Scotland eastwards. This event was important in establishing the present location of the main north–south watershed in the Highlands, which lies close to the west coast so that almost all the main rivers draining the Scottish mountains flow eastwards.

The Palaeogene igneous events produced the youngest solid rocks that now underlie Scotland's mountains, but after the last volcano became dormant and the last dyke had solidified, the Scottish landscape had still to experience over 50 million years of renewed uplift, weathering and erosion to reach its present form. We shall return to this period in the next chapter.

## Conclusion

As you will now appreciate, the geological evolution of Scotland was complex, with the different terranes experiencing radically different histories, further complicated by the fact that the present landmass of Scotland did not exist as a single entity until near the end of the Caledonian Orogeny, around 420–400 Ma. Much of the present geology of Scotland was formed deep within the crust during this orogeny, as thick accumulations of marine sediments were metamorphosed, folded, faulted, intruded by igneous rocks and uplifted to form the Caledonian Mountains, the roots of which now represent most of Scotland's mountains in the Northern Highlands, Grampian Highlands and Southern Uplands. Only the Hebridean terrane of the far northwest escaped, to form a landscape utterly different from that elsewhere in Scotland, with mountains of sandstone and quartzite rising above a primeval platform of Lewisian gneiss. The effects of the orogeny in the Midland Valley, however, have been overprinted by later events, notably crustal stretching and rifting, the deposition of Old Red Sandstone and Carboniferous sedimentary rocks, and two major episodes of volcanism. The opening of the North Atlantic Ocean and the associated Palaeogene igneous activity represent a relatively recent geological postscript to these events, affecting mainly the Inner Hebrides and parts of the western seaboard. Table 2.1 summarizes in chronological order the key events in Scotland's geological evolution.

Throughout this chapter, reference has been made to Scotland's northward drift. To some extent, this has been inferred from the nature of the rock and fossil record during various geological periods, but a more direct method known as *palaeomagnetism* has also been used to reconstruct the former latitude of Scotland. Some rocks, particularly basalts, contain a form of iron oxide called magnetite. When basalts cool and magnetite crystals form, they become aligned along the Earth's magnetic field. As the angle at which the magnetic field reaches the surface (the magnetic inclination) is zero at the Equator and vertical at the poles, the alignment of the magnetic dipoles of magnetite in basalts reflects that of the latitude at which the rock cooled and crystallized, thus allowing the approximate latitude of the rock at the time of formation to be calculated.

Table 2.1 shows the approximate latitude of Scotland during the Phanerozoic eon. During the previous billion years, Scotland (or, more accurately, those fragments of crust that now underlie Scotland) appear to have drifted southwards from near the Equator at the time the Torridonian sandstones were deposited to near the South Pole, before beginning their gradual journey northwards. Throughout much of the Phanerozoic eon,

Scotland lay in low latitudes and experienced a much warmer (often tropical or equatorial) climate than now, but as we drifted into more northerly latitudes (>50°N) during the Neogene period (23.0–2.6 Ma) the global climate began the cool, preparing the stage for the entry of a new factor that was to profoundly influence Scotland's mountain scenery: ice.

Before we consider the nature and effects of the Ice Age in Scotland, however, we turn now to examine the influence of different rock types on mountain relief, examine the events of the 55 million years that have elapsed since the end of Palaeogene igneous activity, and speculate as to what Scotland looked like before the first glaciers began to carve their way through the Scottish landscape.

**Table 2.1** Key events in the geological evolution of Scotland. Latitudes are approximate.

| Period | Age (Ma) | Latitude | Environments and key events |
|---|---|---|---|
| Quaternary | 2.6–0 | ~57°N | Ice age: growth and decay of glaciers and ice sheets. |
| Neogene | 23.0–2.6 | 50–57°N | Initially warm humid climate, cooling after ~10 Ma to temperate humid climate. Phases of uplift, weathering and erosion. |
| Palaeogene | 65.5–23.0 | 45–50°N | Warm humid climate. Opening of Atlantic Ocean. 61–55 Ma: eruption of basalt lavas and intrusion of igneous complexes and dykes of the Hebridean Igneous Province. Uplift of Scotland and tilting to the east. Deep weathering and erosion of uplifted terrain. |
| Cretaceous | 146–65.5 | 40–45°N | Warm humid climate. Shallow seas covered much of lowland Scotland. Continued opening of North Sea Basin. |
| Jurassic | 200–146 | 30–40°N | Warm humid climate. Rising seas deposited marine sediments on low ground. Initial rifting and formation of North Sea Basin. |
| Triassic | 251–200 | 20–30°N | Hot semi-arid environment. Deserts form around Highland margins. Continued deposition of New Red Sandstones. |
| Permian | 299–251 | 10–20°N | Hot arid environment. Scotland lay in continental interior. Deposition of New Red Sandstones in some lowland areas. Intrusion of early Permian sills in the Midland Valley. |
| Carboniferous | 359–299 | 5°S–10°N | Humid equatorial climate. Continued erosion of the Caledonian Mountains. Rifting, subsidence and flooding of the Midland Valley resulted in deposition of limestones, sandstones, siltstones and coal deposits. Eruption of Midland Valley volcanoes and intrusion of the Midland Valley Sill. |
| Devonian | 416–359 | 5–10°S | Hot semi-arid climate. End of Caledonian Orogeny ~390 Ma. Erosion of the Caledonian Mountains and deposition of Old Red Sandstone sediments along margins of the Midland Valley and in the Orcadian Basin. Eruption of Midland Valley volcanoes (Ochil, Sidlaw and Pentland Hills) ~415–410 Ma. |
| Silurian | 444–416 | 10–15°S | Hot semi-arid climate. Closure of Iapetus Ocean. Scandian orogenic event (435–425 Ma): renewed uplift and erosion of the Caledonian mountains, formation of the Moine Thrust Zone and folding of Southern Uplands accretionary complex. ~425–400 Ma: intrusion of 'newer' Grampian granites and eruption of Lorne volcanoes; uniting of the Scottish terranes and connection to England along the Iapetus Suture Zone. |
| Ordovician | 482–444 | 15–30°S | ~480–465 Ma: Grampian orogenic event causes metamorphism, folding and thrusting of Dalradian rocks and intrusion of granites and gabbros in the Grampian terrane. ~470–420 Ma: deposition of greywackes and shales of the Southern Uplands terrane. |
| Cambrian | 542–482 | ~30°S | Iapetus Ocean reaches maximum extent. Deposition of Cambro-Ordovician sediments in NW Scotland. Onset of Caledonian Orogeny ~490 Ma. |
| Precambrian | > 542 | | 600–510 Ma: opening and expansion of the Iapetus Ocean. 730–480 Ma: deposition of Dalradian sediments. 980–870 Ma: deposition Moine sediments. 1200–800 Ma: deposition of Torridonian sediments. 1400–1200 Ma: Lewisian basement uplifted. 3200–2800 Ma: formation of most Lewisian rocks. |

# Chapter 3

# Rocks, relief and the preglacial landscape

## Introduction

In the last chapter we journeyed through time to explore the origins of the rocks that underlie Scotland. In this chapter we consider a rather different question: how and when did the major landscape features of Scotland develop? We have seen that some landscape elements were established by ~400 Ma, such as the Great Glen and main Grampian watershed, and the rifted margins of the Midland Valley certainly existed by ~300 Ma. However, most landscape features in Scotland – uplands and lowlands, mountains and valleys, plateaux and basins – formed much more recently, during the 65.5 million years of the Cenozoic era. The landscape features of Scotland are therefore very much younger than the underlying rocks. To understand landscape evolution during the Cenozoic we need to consider two interrelated questions: first, to what extent have the contrasts between different rock types determined the evolution of Scotland's mountain scenery? And second, how did Scotland's mountain landscapes evolve over the ~55 Ma that separated the end of Palaeocene volcanic activity from the onset of the Quaternary Ice Age? To answer these questions we'll consider first some principles relating landscape evolution to geology, then the relationships between rocks and relief in each of Scotland's terranes. The chapter concludes with a summary of current understanding regarding how the Scottish Highlands evolved during the Cenozoic era, and a glimpse of what the landscape looked like three million years ago, before the coming of the glaciers.

Two concepts underpin the way in which relief and geology are related. The first is *differential weathering and erosion*: if two adjacent rock types of differing resistance experience an identical history of weathering and erosion, the more resistant rock will be left upstanding relative to its less resistant neighbour. This is beautifully illustrated by the volcanic plugs of the Midland Valley, which have been attacked by the same agents of weathering and erosion as the surrounding weaker sedimentary

rocks but now stand proud of the adjacent terrain as steep-sided hills. It is not possible to compile a definitive scale comparing the resistance of different rock types to weathering and erosion, but some broad principles apply in most cases. As the above example suggests, igneous rocks tend to be more resistant than sedimentary rocks. This is because as igneous rocks cool, the mineral crystals become tightly locked together, whereas the grains of sedimentary rocks are more loosely packed and are cemented during lithification, so their strength is determined by that of the mineral cement. A second general proposition is that coarse-grained sedimentary rocks (conglomerates and coarse sandstones) tend to be more resistant than fine-grained sedimentary rocks such as shales and mudrocks. Finally, it is often the case that rocks formed by high-grade metamorphism, notably gneiss, tend to be more resistant than those formed by lower-grade metamorphism, such as slates, phyllites and some mica-schists.

One problem in assessing the erosional resistance of particular rocks is that terms such as 'schist' and 'sandstone' embody rocks of widely varying character. Rocks identified as schists, for example, incorporate both tough, silica-rich granular rocks and also weaker, fissile mica-schists that tend to cleave along foliation planes. Similarly, some sandstones, such as the Torridonian sandstones, have proved much more resistant than others, such as the Carboniferous sandstones of the Midland Valley. Bedrock structure may also cause variation in susceptibility to erosion. Folding, faulting, crushing and jointing greatly influence the intrinsic strength of a rock. A densely jointed granite, for example, is much more vulnerable to weathering and erosion than a granite with low joint density. A further problem is that rock resistance may depend on climatic regime: quartzite, for example, is an exceptionally strong, resistant rock but is vulnerable to breakdown by frost action in cold climates. Finally, chemical weathering of rocks can radically affect their strength. Chemical alteration of biotite in some granites, for example, loosens the framework of such rocks so that they become much more vulnerable to erosion.

The second factor that determines the nature of the relationship between rocks and relief is how slopes and plateaux evolve over millions of years. Geomorphologists have identified three main types of landscape evolution. The first, which is characteristic of resistant horizontal or gently dipping rocks, is *slope retreat*, whereby the hillslopes bordering an uplifted block of ground gradually recede without much change in hillslope gradient (Figs 3.1a and 3.1b). This also occurs where a resistant *caprock* such as a lava flow or sill overlies weaker rocks, protecting them from erosion (Fig. 3.1c). Where resistant and weak rock units alternate, slope retreat forms stepped slopes (Fig. 3.1d). In all of these cases, a low-gradient footslope called a *pediment* progressively forms at the expense of the original uplifted block and eventual coalescence of pediments forms a *pediplain* of low relief, sometimes interrupted by residual hills that represent fragments of the original uplifted block. Where uplifted rocks are uniformly weak, however, hillslopes may evolve by *slope decline*, a gradual decrease in hillslope gradients (Fig. 3.1e) until ultimately a low-gradient plain called a *peneplain* develops. The third possibility is that

an uplifted block experiences deep chemical weathering of bedrock, forming a cover of weathered rock called *saprolite* (Fig. 3.1f). This can occur on all rock types and involves the chemical alteration of particular minerals so that the remaining unaltered minerals form a coarse sandy material called *gruss*, in which are embedded residual rounded blocks of coherent rock called *corestones* (Fig. 3.2). More advanced chemical weathering may lead to pervasive alteration of minerals and the formation of *clayey gruss*, a sticky, clay-rich saprolite. Saprolite covers may subsequently be removed by erosion, revealing an underlying bedrock surface called an *etch surface* (Fig. 3.1h). Because the depth of chemical weathering is variable and often dependent on rock composition and joint density (high joint density favours access by water, the essential ingredient of chemical weathering), the resulting etch surface may be uneven and sometimes interrupted by bedrock knobs, or towers of intact bedrock called *tors*.

These three modes of landscape evolution represent end-members of a continuum of possible evolutionary pathways. They may operate concurrently, depending on structural configuration (Fig. 3.1g), or consecutively; for example, if a period of warm humid climate that favours deep chemical weathering succeeds an arid climate under which chemical weathering is negligible. Uplift of

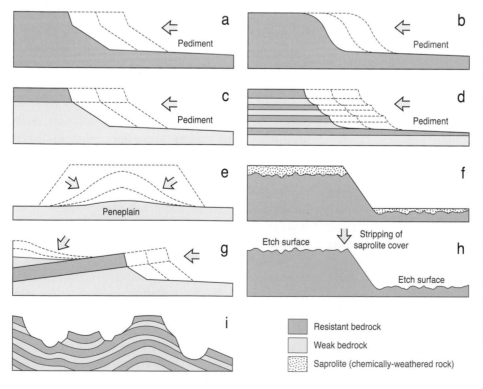

**Figure 3.1** Types of landscape evolution. (**a**) Slope retreat of resistant rock with cliff face. (**b**) Slope retreat of resistant rock without a cliff. (**c**) Slope retreat of a cap of resistant rock overlying weak rock. (**d**) Stepped slope retreat of alternating resistant and weak beds. (**e**) Slope decline on weak rock. (**f**) Development of saprolite cover. (**g**) Combined operation of slope retreat and slope decline. (**h**) Stripping of saprolite cover to form etch surfaces. (**i**) Folding of alternating resistant and weak rocks forms complex topography with the resistant rocks forming higher ground.

**Figure 3.2** Weathered granite in Hill of Longhaven Quarry near Peterhead, northeast Scotland. Rounded corestones occur near the top of the exposure (photograph by John Gordon).

folded rocks introduces complications to the situations depicted in Figure 3.1a–h, particularly when the folds include alternating resistant and weak rocks, so that the broad pattern of folding is inherited by the underlying terrain (Fig. 3.1i). Faults also form zones of weakness within rock masses, as movement along faults results in the formation of a deep belt of crushed rock referred to as *fault breccia.*

All of the scenarios depicted in Figure 3.1 assume no further uplift, but, as we have seen, long-term erosion and removal of rock material leads to renewed isostatic uplift of crustal blocks, either along faults or through crustal warping or tilting. Such uplift is often pulsed: periods of long-term stability may be interrupted by episodes of uplift, during which rivers cut down into pediplains, peneplains and etch surfaces, and the processes of slope retreat, slope decline and/or deep weathering are renewed at a lower level. The result of repeated episodes of pulsed uplift is the formation of a series of *palaeosurfaces* (or *erosion surfaces*), typically separated by scarps up to a few hundred metres high. Several

palaeosurfaces have been identified in the Highlands, particularly in the eastern Grampians, and suggest that over the last 60 Ma the Highlands experienced intervals of relative tectonic stability interrupted by uplift, river incision and formation of younger palaeosurfaces that have expanded at the expense of their higher, older antecedents. Before we explore the ways in which Scottish mountains evolved over this period, it is useful first to consider the overall configuration of the Scottish landscape and how rock type has influenced relief in each of the terranes introduced in the last chapter.

## Structural grain: the ripples of orogeny

A glance at a relief map of Scotland (Fig. 3.3) shows that many mountains, valleys, fjords and firths follow a northeast to southwest alignment known as *structural grain.* This is weaker in the Northern Highlands, where the main mountain ridges are aligned ENE–WSW or east–west, and replaced in Assynt and Sutherland by a distinct northwest to southeast alignment of relief.

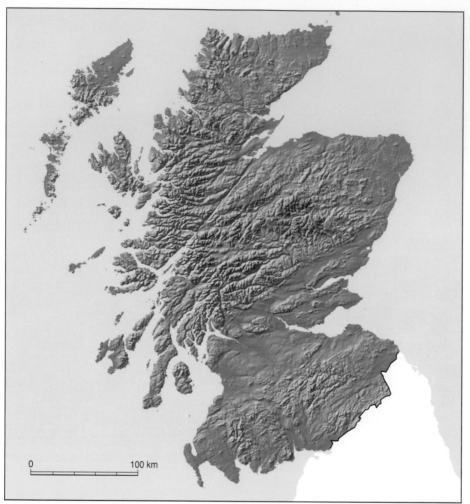

**Figure 3.3** Digital elevation model of Scotland's relief highlights the dominant northeast–southwest structural grain inherited from the Caledonian Orogeny.

0        100 km

The dominant northeast–southwest grain of much of Scotland reflects two factors. First, most major faults are aligned in this direction (Fig. 3.4). The majority of these were activated during the Caledonian Orogeny and represent strike-slip (horizontal) crustal movements as the terranes settled into their final positions, differential (vertical) displacement of crustal blocks, or, in the Southern Uplands, faulting associated with the stacking of the accretionary complex. The zone of crushed rock (fault breccia) associated with faults has been preferentially eroded by rivers and particularly by glacier ice. The Great Glen, for example, is underlain by a belt of fault breccia 1.0–1.5 km wide that has been scooped out by successive glaciers to form Lochs Lochy, Oich and Ness. Similarly, Loch Laidon, Loch Ericht, Glen Truim and upper Strathspey follow the line of the Ericht–Laidon Fault, which extends from Argyll to Nethybridge. In the far north, the northwest–southeast grain is partly determined by older faults, such as the Loch Maree Fault that crosses Wester Ross and the

Loch Shin Fault that extends from the Dornoch Firth to the Moine Thrust Zone. Both form belts of weakness that have been excavated by glacial erosion to form the basins now occupied by the long, narrow lochs that give these faults their names.

The second cause of structural grain is the dominant direction of folding of rocks during the Caledonian Orogeny. Broadly speaking, compression from the southeast has tended to form folds aligned northeast–southwest, particularly in the Grampian Highlands and Southern Uplands. Although there is no simple relationship between the original 'upfolds' (*anticlines*) and high ground, subsequent erosion of folds has juxtaposed relatively resistant and less resistant rocks along northeast–southwest axes, enhancing the structural grain.

The relationship between structure and relief is imperfect, however, and valleys that cut across the dominant structural grain are fairly common. In some cases, this discordance results from *superimposed drainage*: major rivers draining uplifted ground may have

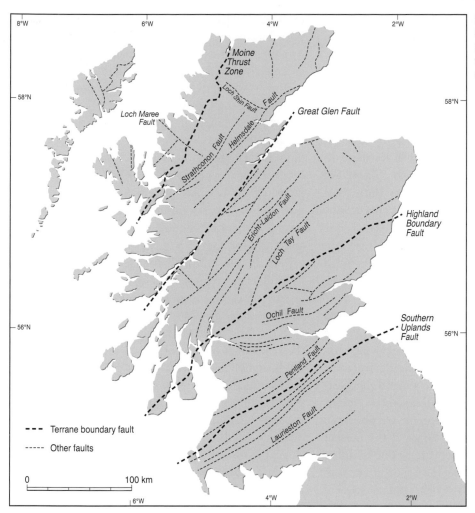

**Figure 3.4** Major faults in Scotland. Those aligned northeast–southwest were formed during the Caledonian Orogeny.

followed the original regional gradient of the uplifted terrain and cut down across the structural trend. The River Tay upstream from Perth, for example, flows southeast across the grain of the Grampians. A further complication is *glacial breaching* (Chapter 5), caused by headward extension of valleys by glacial erosion, forming through valleys. This has affected much of the Northern Highlands, creating west–east aligned valleys, and the long trough now occupied by Loch Lomond probably represents a superimposed drainage routeway deepened and extended by glacial breaching.

## Rocks and relief
### The Hebridean terrane

As outlined in Chapter 2, the mountains of the Hebridean terrane comprise three groups of rocks of different age, origin and character: the Lewisian gneisses that form the crustal basement of Scotland (~3125–1700 Ma), the Torridonian sandstones (~1200–800 Ma) and the

Cambrian quartzites (~530–490 Ma). These three groups are separated by unconformities, implying that the Lewisian Gneisses experienced prolonged erosion before the deposition of the Torridonian sandstones, which in turn were bevelled by erosion before the deposition of the quartzite.

The Lewisian gneiss landscape is the oldest in Britain, a primeval realm that has been exhumed from under the Torridonian sandstones. On the mainland, isolated sandstone mountains such as Suilven and Stac Pollaidh constitute the remnants of the Torridonian cover rocks (Fig. 1.1). Much of the Lewisian landscape is low-lying, but in places Lewisian relief can be seen under the sandstones, as at the foot of Slioch, where the view from across Loch Maree shows ancient Lewisian hills and intervening valleys underlying Torridonian strata. On the mainland, the Lewisian mountains reach their apogee on A'Maighdean (967 m), the oldest of all the Munros, where a cap of Torridonian sandstone meets gneiss near the summit and it is possible to stroll across a billion years from a pebbly Torridonian riverbed onto the tortured Lewisian rocks. Other summits underlain by Lewisian rocks north of Loch Maree include Ben Lair and Beinn a'Chàisgein Mór.

On the Outer Hebrides, the mountains of north Harris, such as Clisham (799 m) and Uisgnaval Mór (729 m) are probably the much-eroded remnants of an uplifted block, though the incorporation of resistant granite intrusions may have reinforced their resistance to erosion. This is probably also the case for the Uig Hills of southwest Lewis, and in south Harris, where the highest summit, Roineabhal (460 m) is underlain by anorthosite, a resistant, coarse-grained, felspar-rich igneous rock. The Lewisian mountains of South Uist, such as Beinn Mhór (620 m) and Hecla (606 m) are the remnants of a tilted block uplifted west of the Minch Fault, which runs along the eastern margin of the Uists.

We do not know how the Lewisian landscape evolved after its emergence (~1400–1200 Ma), though the Lewisian mountains may constitute the surviving remnants of prolonged slope retreat operating from the margins of uplifted blocks. Recent research suggests that low-lying Lewisian terrain subsequently evolved as an etch surface, as pockets of saprolite are preserved in deep fractures. The irregular surface of the present Lewisian lowlands (Fig. 1.1) seems to be due to differential chemical weathering during the Cenozoic era, with deeper weathering along fracture belts, and subsequent stripping of saprolite cover by glacial erosion.

Not all Torridonian rocks form high ground, but those that do underlie some of the most imposing mountains in Scotland. The arkosic sandstone strata of the Torridon Group are amongst the most resistant sedimentary rocks in Britain, are horizontal to gently dipping, and rise grandly towards domed summits or pinnacled ridges along a belt from Assynt to the Kyleakin Hills of Skye. North of Ullapool, the Torridonian mountains form isolated peaks (*inselbergs*) that rise above the Lewisian basement. Stac Pollaidh (613 m), Suilven (731 m) Canisp (847 m) and Quinag (808 m) represent the lonely remnants of Torridonian cover rocks in this area. These residual mountains may have been initially isolated by dissection and slope retreat in the long interval following initial uplift of Torridonian rocks, a process that was renewed following Palaeogene uplift (~60–56 Ma). South of Ullapool, Torridonian rocks form the great sandstone mountains of An Teallach (1062 m), Slioch (980 m), Beinn Alligin (985 m) and Liathach (1054 m), and most of

the intervening low ground. In this area stripping of the Torridonian cover rocks has been less complete; on the Applecross peninsula the Lewisian basement remains completely buried under Torridonian sandstones.

The Cambrian quartzites that unconformably overlie both the Lewisian and Torridonian rocks form a narrow belt along the eastern margin of the Hebridean terrane. In the far north, these resistant rocks form the scree-mantled mountains of Cranstackie (800 m), Foinaven (908 m) and Arkle (787 m), all of which rise proud of the Lewisian basement. Farther south, the quartzites form eastward-dipping resistant caprocks on several Torridonian mountains. The summits of Canisp and Cùl Mór have tiny quartzite caps, and the three eastern summits of An Teallach are crowned by quartzite outliers that represent the upslope continuation of a quartzite escarpment on low ground east of the massif. At the margins of these caprocks it is possible to stand with one foot on the basal quartzite and the other on the sandstones on which it rests, thus straddling hundreds of millions of years of lost geological time (Figs 2.4d and 2.10). On Beinn Eighe above Glen Torridon the quartzite cap thickens eastwards, following the dip of the underlying unconformity, but rugged screes mask the quartzite–sandstone boundary. Along the Moine Thrust Zone, the orderly sequence of Lewisian–Torridonian–Cambrian rocks has been disturbed by thrust faulting and folding. Nowhere is this more evident than at Ben More Assynt (998 m), the ascent of which takes you over stacked Cambro-Ordovician rocks that are succeeded upwards by 'thrust' Torridonian and Lewisian rocks, to reach a quartzite cap on the highest ground. This mountain has a claim to be the most geologically surprising of all those in the Hebridean terrane.

## The Northern Highlands terrane

Geologists have described the Moine rocks that underlie most of the Northern Highlands as 'monotonous', as they lack the variety of the rocks in adjacent terranes, and it is difficult to attribute the topography of mountains formed of Moine rocks to either differential erosion or underlying structure. In part this is because most of the Moine rocks are schists, and in part it reflects an apparent lack of structural control: mountains over 900 m high occur where the schists are vertical, dipping, or intensely folded. The 'Lewisianoid' gneiss inliers within the Moine rocks also underlie both high and low ground. A few major valleys follow faults (notably the Loch Shin fault), but south of the Fannich Mountains the terrain consists of broadly east–west-aligned mountain ridges separated by valleys that are often occupied by long, narrow lochs or fjords: Lochs Fannich, Monar, Mullardoch, Affric, Cluanie, Quoich, Arkaig and Morar, for example, and fjords such as Loch Hourn, Loch Nevis and Loch Sunart. Some of these east–west valleys follow faults, but others cut across major northeast–southwest trending fault lines. The east–west alignment of major valleys and intervening mountain ridges probably reflects a three-stage history: (1) development of drainage towards the Moray Firth or Great Glen during the Devonian period; (2) Palaeogene tilting of the mountain block eastwards, so that the preglacial watershed lay near the west coast and rivers incised long valleys that drain to the Moray Firth and Great Glen; then (3) deepening and headward extension of these valleys by glacial erosion and glacial breaching of the main north–south watershed, as described in Chapter 5.

Although many mountains in the east–west trending ridges of the Northern Highlands have conical summits linked by narrow ridges, others support broad undulating plateaux, often truncated by corrie headwalls (Fig. 3.5). The summit plateaux of Maoile Lunndaidh (1007 m)

in the East Monar hills and Beinn Fhada (1032 m) in Kintail, for example, both extend across ~1 km² and are persuasive evidence for the former existence of a high-level palaeosurface that has been so extensively dissected by rivers, glaciers and landslides that only isolated fragments survive. Supporting this idea is the phenomenon of *accordant summits*: the highest peaks within each of the east–west trending ridges north of Loch Arkaig fall within a narrow altitudinal range (1035–1183 m) and summit altitudes of 900–1100 m are legion. The palaeosurface represented by high plateau fragments and accordant summits constitutes only the oldest palaeosurface of the Northern Highlands; it has been suggested that others form broad benches, low plateaux, basins and plains within and surrounding the mountains.

Perhaps more (geomorphologically) interesting than the 'monotonous' Moines are the isolated mountains of Caithness, east Sutherland and Easter Ross. Some of these, such as Beinn Dhorain (628 m) and Carn Chuinneag (838 m) are underlain by resistant granite intrusions, and Ben Loyal in the far north is underlain by syenite, a tough feldspar-rich igneous rock. Scaraben

**Figure 3.5** Deep corries indented into a palaeosurface fragment (right) on Carn Ghluasaid (957 m), Cluanie. Dipping Moine schists in the foreground.

(626 m) in southern Caithness is an ancient feature, a ridge of Moine quartzite skirted with cemented screes of Devonian age. Its neighbours Morven (706 m) and Smean (509 m) are inselbergs of resistant Devonian conglomerate, and even uplifted blocks of Devonian sandstone locally form high ground, as at Ben Griam Mór (590 m) in east Sutherland and Meall Fuar-mhonaidh, which rises to 696 m above the western shore of Loch Ness.

## The Grampian Highlands terrane

Although the overall pattern of relief in the Grampian Highlands reflects the northeast–southwest structural grain and has been influenced by differential uplift and crustal warping after ~55 Ma, several relationships are evident between high ground in the Grampians and the underlying bedrock. Quartzites have proved particularly resistant, and underlie the Grey Corries ridge in Lochaber, several of the highest peaks of the neighbouring Mamores (Fig. 3.6), and the isolated mountains of Schiehallion and Beinn a'Ghlo in Perthshire. The scree-covered domes of the Paps of Jura are also composed of quartzite and may represent inselbergs

formed by gradual retreat of slopes from the margins of an uplifted quartzite block. Similarly, a belt of tough coarse-grained quartzites ('grits') and quartz-rich schist underlies the mountains of the Highland boundary, such as the Arrochar Hills, Ben Lomond, Ben Ledi and Ben Vorlich. In the mountains around Glen Shee, quartzite tends to underlie the ridges and schist in the intervening valleys, though the relationship is imperfect: a band of pelitic schists underlies the high ground between Glas Maol (1068 m) and Cairn of Claise (1064 m).

Most of the granites intruded into the Dalradian metamorphic rocks of the Grampians form high ground, notably those that form the Cairngorms, Lochnagar and the mountains around Loch Etive, as well as a number of lower, more isolated hills in the northeast Grampians such as Ben Rinnes, Mount Battock, Hill of Fare and Bennachie. There is one striking exception: the Rannoch Moor granite forms a wide, open basin, bisected by the Ericht–Laidon Fault and flanked by mountains of schists and quartzites. This apparent anomaly has been explained by deep chemical weathering of the biotite-rich granite that underlies Rannoch Moor and removal of the weathered rock by erosion. Moreover, in some

**Figure 3.6** Contrasting rocks of the Mamores, western Grampians. In the foreground, the Dalradian quartzite beds of Stob Bàn (999 m) have been exposed by a landslide. In the background are the pink granitic rocks that underlie Mullach nan Coirean (939 m).

areas there is little or no topographic distinction between granites and adjacent metamorphic rocks. On the rolling plateau of the Monadhliath Mountains, for example, granites and schists alike rise to altitudes of 800–940 m, with no evidence of differential erosion.

Of particular interest are the igneous rocks of Scotland's highest mountain, Ben Nevis (1343 m), which rises conspicuously above all adjacent summits. The Ben Nevis complex consists of a granite body roughly 7 km in diameter. Inset within the granite mass is a 2.5 km wide and 600 m thick plug of resistant lavas and *agglomerate* (lava blocks contained within a tough, welded matrix of volcanic ash) that now forms the highest parts of Ben Nevis. This unusual juxtaposition of volcanic and plutonic rocks resulted from a phenomenon known as *cauldron subsidence*: the roof of the magma chamber that fed the Ben Nevis volcano subsided into the underlying magma causing the volcanic rocks to sink into the molten granite. A similar subsidence occurred around Glen Coe, where movement along faults trapped an underlying magma body, causing violent eruptions at fault intersections and the progressive accumulation of a pile of volcanic rocks (andesite and rhyolite lavas, and tough welded ash deposits termed *ignimbrite*) up to 500 m thick. These volcanic rocks are enclosed within granites and schists, and have resisted subsequent erosion more successfully than their neighbours so that they now form high ground, notably Bidean nam Bian (1150 m) and the Three Sisters of Glen Coe, which are largely composed of andesitic and rhyolitic lavas, sills and ignimbrite sheets.

Like the mountains of the Northern Highlands, those of the western Grampians tend to form accordant summits, mainly within the range 900–1100 m, and a few support plateau fragments that are probably remnants of strongly dissected palaeosurfaces. On the Monadhliath Mountains and in the central and eastern Grampians, however, high-level palaeosurfaces are much more extensive, forming wide, rolling plateaux interrupted by glacially eroded troughs (Fig. 3.7). The Gaick plateau east of Drumochter Pass, for example, forms an undulating plateau above 800 m, and east of Lochnagar another broad plateau descends gradually eastwards from ~800 m to ~500 m. Such plateaux represent palaeosurfaces produced by a combination of differential uplift and deep weathering during the Cenozoic era, as outlined later in this chapter. Differential uplift also

**Figure 3.7** Part of the Eastern Grampian palaeosurface. The summit on the right is Driesh (947 m), and the trough on the left is Glen Clova (photograph by Martin Kirkbride).

explains a geological conundrum: the rocks that form the Grampian mountains are similar to those that underlie the Buchan Plain, an extensive area of low ground in northeast Scotland that rarely exceeds 200 m in altitude. This contrast in relief cannot be explained by differences in rock types, and, as outlined below, reflects contrasts in the Cenozoic tectonic history of different parts of the Grampian Highlands terrane.

### The Southern Uplands terrane

In Chapter 2, we saw that the Southern Uplands form an accretionary complex, dominated by tough greywackes interbedded with shales. This complex was pushed over the continental basement rocks as the Iapetus Ocean closed, forming a high mountain chain and uniting Scotland with England at the Iapetus Suture Zone. The compression caused by this event produced a series of faults aligned northeast–southwest (Fig. 3.4) and rumpled the rocks into a series of tight folds, roughly parallel to the faults, so that the greywackes often dip steeply to the southeast. These structures have been eroded to form a series of dissected plateaux aligned northeast–southwest, characterized by rolling hills 450–600 m high, notably the Moorfoot Hills, Lammermuir Hills and Cheviot Hills. In the west, three discordant valleys cut across the structural grain. Of these, Nithsdale and Annandale are partly floored by Permian sandstones, suggesting that both valleys represent ancient drainage routeways superimposed across the structural trend as the Southern Uplands were uplifted. The third, between Dalmellington and Castle Douglas, follows a fault.

The highest mountains of the Southern Uplands are all related to the underlying geology. The Tweedsmuir and Moffat Hills, which culminate in Hart Fell (808 m), White Coomb (821 m) and Broad Law (840 m), are underlain by exceptionally thick beds of gritty resistant greywacke. Most of the granite intrusions in the southwest have also proved relatively resistant to erosion, and underlie Cairnsmore of Carsphairn (797 m), Cairnsmore of Fleet (711 m) and the isolated mass of Criffel (569 m) that rises boldly from the shores of the Solway Firth. The Loch Doon granite intrusion, however, represents a fascinating geological anomaly, for much of the granite forms low ground surrounded by an amphitheatre of high, rugged mountains culminating in Merrick (843 m) and Corserine (814 m). This ring of mountains is underlain by tough metamorphosed greywackes that were 'baked' as the molten granite pluton ascended (a classic example of contact metamorphism), and now form a *metamorphic*

*aureole* that surrounds a basin eroded in granite that was weakened by deep chemical weathering.

### The Midland Valley terrane

'Midland Valley' is a misnomer, as that part of Scotland lying between the Highland Boundary Fault and the Southern Uplands Fault is studded with hills and uplands formed by the stacked lava flows, volcanic plugs and sills associated with Devonian and Carboniferous igneous activity. The Midland Valley demonstrates *par excellence* the concept of differential erosion, which has worn down the sedimentary rocks but left the igneous rocks standing proud. Holyrood Park in Edinburgh provides a microcosm of this relationship: above the adjacent sandstones and shales the high ground is formed by the twin volcanic plugs of Arthur's Seat (251 m), the lavas of adjacent Whinny Hill and the bold prow of the Salisbury Crags sill that rears above the Scottish parliament building. On a wider canvas, the 90 or so volcanic plugs of the Midland Valley (which include Dumbarton Rock, Dumgoyne, Edinburgh Castle Rock, Largo Law, North Berwick Law and the Bass Rock) reach their greatest elevation in West Lomond Hill (522 m), and the great arc of Carboniferous plateau basalts that partly encircles Glasgow culminates in Earl's Seat (578 m) in the Campsie Fells. The Midland Valley sill that is intruded into the Carboniferous sedimentary rocks around the inner Firth of Forth is ~200 m thick and forms the Lomond Hills scarp in Fife. Much of the highest ground within the Midland Valley, however, is that associated with the Devonian lavas that form the Ochil Hills, which culminate in Ben Cleuch (721 m) and King's Seat Hill (648 m), and the Pentland Hills, which reach their highest point at Scald Law (579 m).

By contrast, the sedimentary rocks of the Midland Valley rarely rise above 200 m, with two interesting exceptions. South of Strathaven, a fault-bounded inlier of resistant greywackes underlies the Glengavel Hills, which culminate at Nutberry Hill (522 m) and Hagshaw Hill (488 m); and along the margin of the Highland Boundary Fault, the conglomerates that represent the fragmented remains of Old Red Sandstone alluvial fans have resisted erosion to form Uamh Bheag (664 m) and neighbouring Beinn Odhar (631 m).

Faulting has played a key role in forming steep scarps in the Midland Valley. Faults define the steep south face of the Campsies, the southeast boundary of the Pentlands, the Lomond scarp, the northern margin of the Fintry Hills and the edge of the Sidlaw Hills where they rise above the Carse of Gowrie east of Perth. Most

**Figure 3.8** The Devonian lavas of the Ochil Hills. The steep southern face of the Ochils represents a fault scarp that follows the Ochil Fault, which extends eastwards from near Stirling across Fife.

impressive of all is the 350 m high fault scarp that defines the southern margin of the Ochils between Stirling and Dollar. This impressive wall of stacked lava flows (Fig. 3.8) forcibly demonstrates that use of the description 'Midland Valley' to describe the landscape of central Scotland fails to do justice to this region of strikingly varied relief.

### The Hebrides Igneous Province

The Palaeogene basalt lavas that underlie much of Mull, northern Skye and southern Morvern tend to form a low, monotonous stepped landscape, though they underlie high ground in northern Skye in the form of MacLeod's Tables (488 m and 468 m) and the Trotternish escarpment, where westward tilting of the lava pile has created an east-facing scarp that culminates in The Storr (719 m) and seven other peaks over 500 m (Fig. 2.4c). The bold monolith of An Sgùrr (393 m) on the island of Eigg represents extremely resistant lava called *pitchstone* that was erupted around 52 Ma and originally infilled a canyon within older basalt lava flows that have now been largely removed by erosion. The basalt lavas reach their highest point at the summit of Ben More (966 m) on Mull, where an uplifted cap of lavas erupted at ~60–58 Ma gives the mountain a claim to be (geologically) the youngest of all the Munros.

The central igneous complexes on Skye, Rum, Mull and northern Arran all form mountains that are dissected by glacial troughs and scalloped by corries, but the nature of the relief is partly conditioned by the underlying rock type. This is strikingly illustrated by the contrast between the gabbro and granite mountains of central Skye. The tough, resistant gabbros of the Cuillin Hills and Blà Bheinn form the most rugged mountains in Scotland and include several summits over 900 m (Fig. 3.9a); the jagged relief of these mountains is emphasized by preferential erosion of the basalt dykes that cut through the gabbro. In contrast, the granite that underlies the adjacent Red Hills forms domed summits and ridges less than 800 m high (Fig 3.9b), suggesting that the granites have been more susceptible to weathering and erosion than their gabbroic neighbours. On Rum, the layered mafic rocks that underlie Askival (812 m) and Hallival (722 m) also form rugged peaks, whereas the western granite hills of the northern Arran complex closely resemble the domed Red Hills of Skye. The eastern granite hills of northern Arran, however, have been extensively modified by glacial erosion and selective weathering of basalt dykes, and now form peaked summits and arêtes that almost rival the Cuillin Hills in terms of topographic ruggedness.

**Figure 3.9 (a)** The Cuillin Hills, Skye, composed mainly of gabbro. The highest peak is Sgùrr a'Ghreadaidh (973 m). **(b)** The granite Red Hills of Skye.

## Cenozoic landscape evolution

As outlined in Chapter 2, some major elements of Scotland's landscape were established by ~400 Ma, such as the main Grampian watershed, the Great Glen, the margins of the Moray Firth basin and various smaller basins in the Highlands. Others, such as the rifted Midland Valley, were in existence by ~300 Ma. By the end of the Cretaceous period, however, much of Scotland was probably reduced by erosion to a lowland landscape of limited relief, implying that most Scottish mountains were uplifted to their present altitudes during the Cenozoic era.

The Cenozoic timeline is divided into three geological periods (Palaeogene, Neogene and Quaternary), each of which is further subdivided into geological epochs of unequal duration (Fig. 1.6). The Palaeogene is subdivided into the Palaeocene (65.5–55.8 Ma), Eocene (55.8–33.9 Ma) and Oligocene (33.9–23.0 Ma); the

Neogene comprises the Miocene (23.0–5.3 Ma) and Pliocene (5.3–2.6 Ma); and the Quaternary is subdivided into the Pleistocene (2.6 Ma to 11.7 ka) and Holocene (11.7 ka to the present). Here we are concerned with landscape evolution during the Palaeogene and Neogene periods, during which Scotland's relief began to assume its present form, before the first appearance of glacier ice in Scotland during the Quaternary.

Reconstructing the preglacial events of the Cenozoic era is challenging, in part because of the ravages of later glacial erosion, but also because only tiny fragments of preglacial Cenozoic sediments are preserved on land, so that establishing the timing of events is problematic. Much thicker deposits are preserved offshore, however, and these provide a rough indication of what was happening on land: periods of rapid offshore sediment accumulation suggest that the land was being uplifted and rapidly eroded, whereas periods of reduced sediment accumulation indicate intervals of relative stability.

Most research on preglacial landscape evolution has focused on the terranes north of the Highland Boundary Fault and suggests that four main factors influenced the development of relief prior to the advent of glaciation. First, as noted above, some major landscape elements are inherited features, some of which probably date back to 400 Ma (or in the case of the Lewisian topography, exhumed from under the Torridonian sandstones) very much earlier. These inherited elements were modified but essentially preserved throughout the Cenozoic era. Second, from a starting point of generally low relief at the onset of the Cenozoic, the Highlands have experienced differential tectonic activity: some areas were uplifted more than others, some were tilted, and others were downwarped so that they now form low ground despite being underlain by the same rock types that form adjacent mountains. Third, there is evidence that gradual slope retreat occurred during periods of relative stability, extending basins and lower palaeosurfaces at the expense of higher palaeosurfaces. Finally, throughout much of the Cenozoic era Scotland experienced a subtropical climate, due to global climate warming, so that during the Palaeogene and much of the Neogene warm humid conditions prevailed. As a result, high ground and low ground alike experienced deep chemical weathering and the formation of saprolite covers that have now been largely removed by erosion: the palaeosurfaces and basins of the Highlands are essentially etch surfaces modified to varying degrees by glacial erosion during the Quaternary.

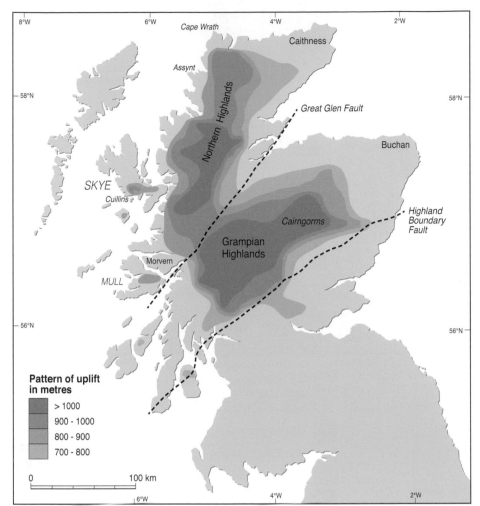

**Figure 3.10** Generalized pattern of Cenozoic uplift (in hundreds of metres) implied by summit altitudes in the Scottish Highlands and Inner Hebrides. This interpretation represents the outcome of several episodes of uplift, tilting and downwarping. Adapted from Hall, A.M. (1991) *Transactions of the Royal Society of Edinburgh: Earth Sciences*, 82, 1–26, with permission from the Royal Society of Edinburgh.

Because the Highlands probably formed a region of generally low relief at the start of the Cenozoic era, the overall pattern of Cenozoic uplift can be reconstructed from present summit altitudes (Fig. 3.10). Major uplift of the Highlands occurred in the mid- to late- Palaeocene (~60–56 Ma) and appears to represent a response to magmatic activity in the Hebridean Igneous Province. The Northern Highlands and Grampian Highlands may have been uplifted by up to 1000 m and tilted as a single rigid block, though with downwarping at its margins, particularly in Caithness and northeast Scotland. Uplift was renewed in the late Oligocene (~27–23 Ma), followed by relative tectonic stability throughout the early- to mid-Miocene (~23–11 Ma), then renewed slow uplift of some areas in the late Miocene and Pliocene (~11–2.6 Ma) that probably continued into the Pleistocene.

The major initial Palaeocene uplift was accompanied by vigorous erosion that stripped most Jurassic and Cretaceous sedimentary cover rocks from the Highlands, exposing the older metamorphic rocks of Lewisian, Moine and Dalradian age. In the northwest Highlands, uplift created a steep westward-facing scarp, now heavily dissected, that extended from Cape Wrath to Skye, and renewed slope retreat further isolated the Torridonian sandstone inselbergs such as Suilven and Quinag. The successive tectonic movements that affected the Highlands were uneven: some areas uplifted in the Palaeocene were tilted or downwarped by later tectonic events. As a result, it is difficult to establish the contemporaneity or otherwise of high-level palaeosurface fragments in the strongly dissected mountains of the Northern Highlands and western Grampians. Only in the eastern Grampians, where much more extensive plateaux are preserved (Fig. 3.7), is it possible to identify with confidence major high-level palaeosurfaces, notably the Eastern Grampian Surface, Monadhliath Surface and Gaick Surface, all of which decline generally northeastwards from 800–900 m to ~500 m at their outer margins (Fig. 3.11).

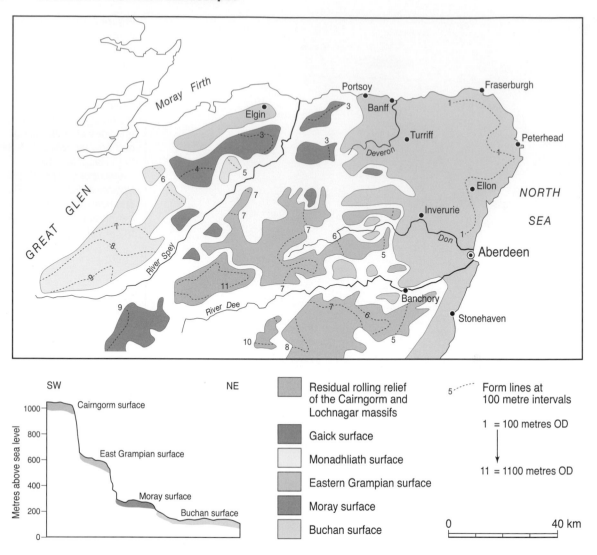

**Figure 3.11** Palaeosurfaces (erosion surfaces) in the Grampian Highlands terrane. Adapted from Hall, A.M. (1991) *Transactions of the Royal Society of Edinburgh: Earth Sciences*, 82, 1–26, with permission from the Royal Society of Edinburgh.

The role of deep chemical weathering in the Cenozoic evolution of Highland relief is attested by the preservation of pockets of saprolite up to 10 m deep on mountains underlain by a range of rock types. Remnants of sandy gruss occur, for example, on the quartzite of Arkle in the far northwest, on the lavas of Skye, high on the granite plateau of the Cairngorms and on the schists of the Gaick plateau. Both sandy and clayey gruss saprolites are more common on the Buchan Surface (Figs 3.2 and 3.11) of northeast Scotland, where they reach depths of up to 50 m, and are even preserved within fissures in the ice-scoured Lewisian rocks of northwest Scotland. The sandy gruss saprolites are thought to have formed no earlier than the Pliocene, but the clayey grusses may reflect prolonged rock weathering dating back to the

Miocene or earlier. These remnant saprolites provide evidence of extensive deep weathering that modified palaeosurfaces at all altitudes. An important implication is that present-day plateaux and palaeosurface fragments in the Highlands are not pristine preglacial surfaces that were uplifted during the Cenozoic but represent etch surfaces from which saprolite covers have been removed by subsequent erosion.

Deep weathering affected some rocks more than others, and in particular was responsible for the formation of basins (relatively low ground partly encircled by steep slopes) during periods of tectonic stability. Such basins appear to have developed through preferential deep weathering of the underlying rock, retreat of surrounding slopes and deepening through erosion,

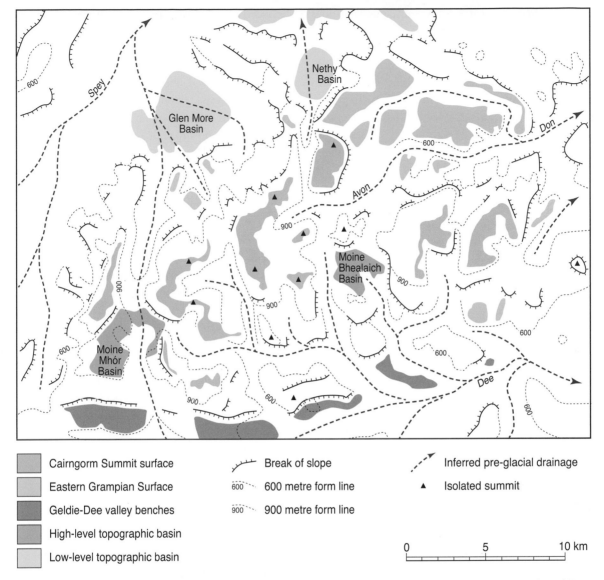

**Figure 3.12** Preglacial landforms of the Cairngorm Mountains. Adapted from Hall, A.M. and Gillespie, M.R. (2016) Fracture controls on valley persistence: the Cairngorms granite pluton, Scotland. *International Journal of Earth Sciences*, 106, 2203–2219. © 2016 The authors.

particularly during periods of uplift. The Rannoch Moor basin is a classic example, formed by deep weathering and erosion of biotite-rich granite; the Inchbae and Naver basins of the Northern Highlands and the Loch Doon basin in the Galloway Hills also owe their formation to a combination of deep weathering, erosion, and retreat of surrounding slopes.

Basins are not restricted to low ground, however, but are locally developed on upland areas. This is illustrated by the preglacial (though subsequently glacially modified) landforms of the Cairngorms. This granite massif supports fragments of two palaeosurfaces separated by a scarp ~200 m high: the Cairngorm Summit Surface, a gently

rolling plateau mainly ~1100–1300 m in elevation, and, in the northeast, the Glen Avon embayment, most of which lies between 750 m and 1000 m and which represents part of the Eastern Grampian Surface (Fig. 3.12). On the margins of the massif are broad valley benches, formed by uplift of the floors of the valleys formerly occupied by the Rivers Geldie and Dee, and upland basins, such as the Glen More basin in the northwest, the Nethy basin in the north, and in the southeast the high-level Mòine Bhealaich basin, most of which lies above 800 m. As Figure 3.12 shows, much of the large-scale relief of the Cairngorms (palaeosurfaces, scarps, valley benches, basins and the main drainage routeways) formed before the onset of Quaternary glaciation. Subsequent episodes of glacial erosion resulted in removal of saprolite covers, deepening and widening of valleys and basins, breaching of watersheds and the formation of corries at plateau margins, but these changes represent modification of large-scale topographic features of much greater antiquity. The Cairngorms, Monadhliath Mountains, Gaick plateau and southeast Grampians constitute the best-preserved Neogene upland landscapes in Scotland, but relict surfaces, scarps and basins of Cenozoic age can be detected across much of the central and eastern Highlands. Only in the western and northwest Highlands are such features rare, as a result of vigorous Palaeogene and Neogene erosion and even more vigorous dissection by Pleistocene glaciers.

## Synthesis

As will be apparent from this chapter, differential weathering and erosion of rocks of contrasting resistance offer only a partial explanation of present relief in Scotland. Although there are clear correspondences between rock type and relatively high ground in the Midland Valley, Hebridean Igneous Province, Caithness and Sutherland, and (to a lesser extent) the Southern Uplands, elsewhere the relationship between lithology and relief is obscure or compromised by exceptions. Within the Hebridean terrane, for example, Lewisian gneiss, Torridonian sandstone and Cambrian quartzite all underlie both mountains and low ground. Similarly, there is no clear relationship between topography and the Moine rocks of the Northern Highlands. In the Grampian Highlands some rock types, such as quartzite, granite and volcanic rocks typically form high ground and have locally resisted weathering and erosion more successfully than neighbouring schists, but there are exceptions, and it is particularly notable that the palaeosurfaces of the eastern Grampians cut across various rock types with little or no change in relief. Similarly, the accordant summits throughout much of the western Highlands fall within a fairly limited altitudinal range irrespective of the underlying rock type, whether it be Torridonian sandstone, Moine schists, quartzite, Dalradian metasediments or granite.

The limited influence of contrasting rock types on the pattern of relief north of the Highland boundary is due in part to *landscape inheritance*, the tendency for mountains and valleys to persist in roughly the same locations over very long timescales. Mountains, ridges and plateaux

are slowly eroded rock masses separated by downcutting rivers, and because the courses of rivers have tended to remain broadly constant during successive periods of uplift, the location of the intervening high ground also remains unchanged, irrespective of rock type. Moreover, Cenozoic tectonic activity has also played a major role (arguably *the* major role) in dictating which parts of the Highlands now form high ground. Uplift of the late Cretaceous landscape of low relief explains the broad accordance of summits of varying lithology across much of the Highlands, and the tendency for palaeosurfaces to extend uninterruptedly across different rock types. Conversely, tilting and downwarping of crustal blocks during the Cenozoic explains why some rocks that underlie mountains are identical to those that underlie low-lying terrain. Most of the Dalradian rocks that form the mountains of the south-east Grampians, for example, are identical to those that underlie the low-lying Buchan surface in northeast Scotland, but the two areas have experienced different tectonic histories. Differences in rock resistance to prolonged weathering and erosion have certainly played a role in determining the distribution of high ground in some parts of the Highlands, but this role has been secondary to the combined influence of landscape inheritance and patterns of Cenozoic tectonic activity, slope retreat and deep weathering.

## Before the ice: the preglacial landscape of Scotland

If we could charter a flight across the Scottish mountains three million years ago, before the advent of the earliest Pleistocene glaciers, most large-scale features of the present landscape would already be evident. The northeast–southwest structural grain of the Grampians and Southern Uplands was already etched out along faults by differential erosion, and we should have no difficulty in identifying the main mountains and massifs. Ben Nevis already rears above the surrounding summits, Schiehallion rises proudly above the adjacent terrain and the granite mass of the Cairngorms stands above the adjacent plateaus. The igneous uplands of the Midland Valley already form high ground overlooking the adjacent less resistant sedimentary rocks, and in the far northwest we should certainly be able to identify Suilven, Stac Pollaidh and their more massive cousins, the mountains of Wester Ross and Torridon. The Palaeogene igneous centres of Skye, Rum and Mull were probably at roughly their present elevation, partly flanked by stepped lava flows.

Throughout Scotland, present-day plateaux, summits, basins, scarps and most drainage patterns were already in existence.

We might also observe, however, that the end-Neogene landscape of Scotland was less rugged than now: slopes were gentler, summits more rounded, cliffs and rock outcrops rarer, and palaeosurfaces less dissected. We would also see that very few lochs were present, because most Scottish lochs occupy rock basins excavated by glacier ice. For a similar reason the deep fjords of the western Highlands would probably have been represented by shallow inlets, and it is likely that some of the islands of the Inner Hebrides, such as Skye and Mull, were attached to the mainland. In the mountains, opportunities for rock climbing must have been few, as corries were absent, and the narrow arêtes that now give such joy to hillwalkers were similarly rare or non-existent. Thus although the main elements of Scottish mountain topography existed before the glaciers, many of the most distinctive Scottish mountain landforms awaited the coming of the ice. The next two chapters outline the characteristics of the Ice Age in Scotland, and how successive episodes of glaciation have modified Scotland's mountain landscapes to their present form.

# Chapter 4

# The Ice Age in Scotland

## Ice ages: an introduction

Throughout much of its 4.54 billion-year history, planet Earth has been much warmer than now: so much warmer that glacier ice has been completely absent, even at the poles. Over timescales of tens of millions of years, however, the planet has several times cooled sufficiently for glacier ice to become widespread over continental landmasses. We know this because ancient glacial deposits called *tillites* appear in the geological record for certain times in the past but are absent for others. The oldest ice age recognized in the geological record, the Huronian, lasted from ~2400 Ma to ~2100 Ma. The Cryogenian Ice Age of ~720–635 Ma probably involved the most extensive cover of glacier ice across the continents; in Scotland this ice age is represented by the Port Askaig Tillite, a rock formation found on Islay and the Garvellach Islands in the Firth of Lorn. Extensive continental ice sheets also formed during the Andean–Saharan Ice Age (~450–420 Ma) and the Permo-Carboniferous Ice Age of ~360–260 Ma. But when we discuss *the* Ice Age, we are usually referring to a much more recent event in Earth history, one that is known as the *Cenozoic Ice Age* or the *Quaternary Ice Age*. The former term refers to the period since glacier ice began to accumulate on Antarctica around 35 million years ago. As we saw in Chapter 3, the *Quaternary period* began much later, at ~2.6 Ma, an age determined by the earliest evidence of glacial conditions extending to mid-latitudes. The Quaternary period incorporates two geological epochs: the *Pleistocene* (2.6 Ma to 11.7 ka) and the *Holocene* (11.7 ka to the present). Many scientists have argued that we have now entered a new epoch, the *Anthropocene*, a view that acknowledges the role of human activity in influencing recent environmental change and some geological processes. There is limited agreement, however, as to when the Anthropocene began, and some geologists consider that this concept should only be employed informally.

The causes of ice ages have stimulated much debate, though it is generally accepted that continental drift has played a part in preconditioning ice-age initiation. Clustering of continental landmasses around the Equator is unlikely to favour the onset of an ice age, but movement of continents into high latitudes permits the growth of ice sheets and ice caps at and near the poles. Continental drift may also have resulted in blocking warm ocean currents from reaching polar areas, encouraging cooling of high latitudes. Similarly, uplift of mountain ranges may have altered atmospheric circulation, diverting warm airmasses. All of these factors favour the development of glacier ice around the poles, but they are inadequate to explain the *global* cooling associated with full ice-age conditions when glacier ice extended over mid-latitude areas. One idea favoured by many scientists is that such global cooling was caused by a slow reduction of atmospheric carbon dioxide, the most abundant of the greenhouse gases that retain heat in the atmosphere. Reduction of atmospheric carbon dioxide concentration has been explained by vigorous weathering and erosion of newly uplifted mountain chains, which is thought to lead to increased storage of carbon in oceanic sediments, thus removing it from the carbon cycle that returns carbon to the atmosphere as carbon dioxide. According to this view, the geologically recent uplift of the Tibetan Plateau and mountain chains such as the Andes, Alps and Himalayas may have been indirectly responsible for gradual atmospheric cooling over the last ~35 Ma, and the more abrupt descent into the Quaternary Ice Age 2.6 million years ago.

## Glacial and interglacial stages

If you consider the duration of the ancient ice ages identified above, it becomes obvious that these lasted tens or even hundreds of millions of years. By contrast the 2.6 million-year duration of the Quaternary Ice Age – *the* Ice Age – appears very brief. This is because we are still in the Ice Age but living in a temperate interval (the Holocene) when glacier ice occupies only about 10% of the Earth's land area, mostly in Antarctica and Greenland. Studies of the pattern of global climate change based on changes in oxygen isotopes in deep-ocean sediment cores and in ice cores extracted from the Antarctic and Greenland Ice Sheets have demonstrated that the Quaternary period witnessed numerous climatic shifts between cold *glacial stages* and intervening warmer *interglacial stages* (Fig. 4.1). The Holocene represents the present interglacial, and in the absence of anthropogenic climate warming the planet would almost inevitably experience a return to glacial conditions at some time in the future. Within glacial and interglacial stages there were shorter intervals of colder conditions that we refer to as *stades or stadials*, and brief intervals of warmer conditions that we call *interstades* or *interstadials*. Towards the end of the most recent glacial stage in Scotland, for example, rapid warming at ~14.7 ka introduced a period of cool temperate climate called the *Lateglacial Interstade*, when summer temperatures initially approached those of today. Climatic deterioration at ~12.9 ka, however, brought a brief return to extremely cold glacial conditions during the *Loch Lomond Stade*, which lasted until ~11.7 ka, when renewed rapid warming ushered in the Holocene.

The causes of glacial–interglacial oscillations during the Quaternary are complex, but are related to the geometry of the relationship between the Earth and the sun, and its implications for the amount of solar radiation received in northern high latitudes (> 55°N) where the great Quaternary ice sheets expanded during glacial stages and contracted during interglacials. This concept is known as the *astronomical theory of climate change*, and involves the interplay of three factors: the eccentricity of the Earth's orbit, which varies from near-circular to an asymmetrical ellipse and back to near-circular over a ~100 ka cycle; orbital precession, which determines at what season the Earth is closest to the sun (at the moment this occurs on January 3rd) and follows a ~26 ka cycle; and the tilt of the Earth's axis relative to the plane of its orbit, currently 23.4°, which varies from 22.1° to 24.5° over a 41 ka cycle. Over millennial timescales, these effects sometimes cancel out, but at other times they reinforce each other to increase or reduce the solar radiation reaching northern high latitudes. In the 1970s, when it became possible to date the ages of Quaternary glacial–interglacial oscillations, it was shown that the timing of these oscillations corresponded with the predictions of the astronomical theory, suggesting that the relationship

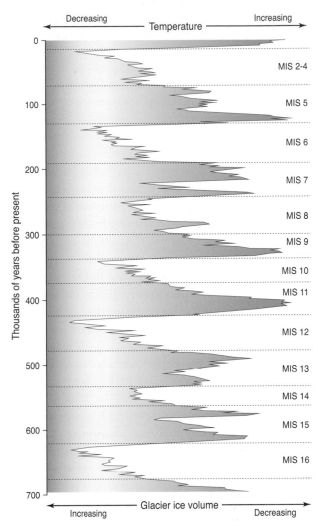

**Figure 4.1** Glacial–interglacial oscillations over the past 0.7 Ma, recorded by changes in the oxygen isotope ratios recorded for marine microfauna in sediment cores from the ocean floor. 'MIS' refers to *marine isotope stage*, a convenient way of identifying glacial stages (even numbers) and interglacial stages (odd numbers). Shorter-term spikes indicate the timing of stadial and interstadial events. During the coldest parts of glacial stages, glaciers occupied up to 30% of the Earth's land surface, whereas during the warmest parts of interglacials they covered ~10%, mainly in Greenland and Antarctica. Glacial–interglacial oscillations occurred throughout the Quaternary but their amplitude increased after ~0.9 Ma.

between the three factors outlined above – orbital eccentricity, orbital precession and axial tilt – represents the pacemaker of the Quaternary Ice Age.

However, the changes in solar radiation received in northern high latitudes due to the interplay of these three factors are apparently too small to have initiated *global* cooling during glacial periods or warming during interglacials. Such radical climate changes appear to depend on feedback effects that are triggered by changes in northern high-latitude incident radiation. One feedback relates to the reflectivity or *albedo* of the Earth's surface. If slight northern hemisphere cooling causes the expansion of ice sheets and summer sea-ice cover in the Arctic, global albedo increases because snow and ice have higher reflectance than land surfaces or open water. An increase in global albedo results in more solar radiation being reflected back out to space instead of warming the land surface, ocean and atmosphere. This in turn induces further cooling, leading to further expansion of snow and ice cover, which increases global albedo further, leading to further cooling and so on. Conversely, as ice sheets and sea-ice cover begin to shrink in response to warming, global albedo is reduced, enhancing further warming. Some scientists have also argued that a reduction in carbon dioxide in the atmosphere initiated or accompanied expansion of the northern hemisphere ice sheets, contributing to enhanced cooling. It is unclear, however, whether changes in atmospheric carbon dioxide concentration represent a response to, rather than a cause of, ice-sheet expansion.

By virtue of its location, Scotland has been particularly susceptible to rapid climate changes at the onset and termination of glacial stages. Scotland occupies a latitudinal range (55°N–61°N), similar to that of southern Alaska, but presently enjoys much milder winters. The reason for this contrast is that warm surface water from the tropics is driven across the North Atlantic Ocean by a surface current, the North Atlantic Drift (or 'Gulf Stream'), warming the seas around northwest Europe. As this warm surface saltwater moves northward and cools, it becomes denser and sinks, returning slowly southwards as deep water within the ocean. There is thus a natural oceanic 'conveyor belt' known as *Atlantic meridional overturning circulation* (AMOC) that is responsible for bringing relatively warm surface waters to the North Atlantic Ocean. Additionally, warm airmasses originating in the tropics periodically engulf Scotland, even in the depths of winter. As a result, Scottish winters are less severe than those of southern Alaska, which is surrounded by cold polar waters and experiences less frequent incursions of warm tropical air.

During glacial stages, however, increased discharge of freshwater (meltwater from expanding ice sheets) into the North Atlantic Ocean disrupted AMOC, and cold polar surface waters advanced southwards, enveloping the British Isles and causing warm airmasses to follow a more southerly trajectory. At the time of the global Last Glacial Maximum (~23 ka), for example, the boundary between cold polar and warm tropical surface water (the *North Atlantic oceanic polar front*) migrated southwards to the latitude of northern Spain. As most weather in Scotland comes from the North Atlantic, replacement of warm surface water by cold polar water and associated southward migration of warm airmasses had a drastic effect, causing air temperatures in Scotland to plummet. Conversely, towards the end of glacial stages, resumption of AMOC and consequent northward retreat of the oceanic polar front resulted in return of warm surface waters to Scotland's coasts and renewed incursions of warm tropical airmasses. These effects caused extremely rapid atmospheric warming around ~14.7 ka, possibly by as much as 1°C per decade.

Scotland therefore occupies a location that has been extremely sensitive to glacial–interglacial or stadial–interstadial climatic oscillations during the Quaternary, as the changes in the temperature of North Atlantic surface water and the associated changes in the trajectory of warm airmasses amplify and accelerate more general global warming and cooling trends. As a result, the successive glaciers and ice sheets that occupied much or all of Scotland responded nimbly to Quaternary climate changes. Most of our understanding of glacier response, however, is limited to the very last of the multiple glaciations that affected Scotland during the Pleistocene, as the most recent Scottish Ice Sheet obliterated much of the evidence for earlier glaciations. Before we consider the history of the last ice sheet and later mountain glaciers in Scotland, however, it is useful to summarize the behaviour of glaciers – how they respond to climate, how they move, how they erode and deposit debris – and the formation and characteristics of some common glacial landforms.

## Glaciers and glaciation
### Classification of glaciers

To many people, the word *glacier* conjures up an image of a broad river of ice moving slowly along the floor of a mountain valley, but glaciers exist in many shapes and sizes. A basic distinction can be drawn between those that are *topographically unconstrained*, in the sense that the direction of ice movement is not determined by the underlying topography, and those that are *topographically constrained*, implying that ice movement is focused along valleys.

Topographically unconstrained glaciers completely bury the underlying landscape, at least in their central parts, and the direction of ice movement is away from the highest point of the ice surface. Such glaciers can therefore move across mountain barriers, though at their periphery, where the ice cover thins, they may become increasingly focused in valleys. Those exceeding 50,000 km² in area are termed *ice sheets*. At present only two ice sheets exist, the Antarctic and Greenland Ice Sheets, but during the later glacial stages of the Pleistocene ice sheets covered all of Scandinavia and much of northern North America and the British Isles. Smaller topographically unconstrained glaciers are called *ice caps* and are much more common at present; examples include the Vatnajökull Ice Cap in Iceland and Jostedalsbreen in Norway.

The surface of ice sheets is often uneven, comprising broad *ice domes* that rise above the general level of the ice surface and act as centres of radial-outwards ice movement, and *ice divides* from which ice moves in two opposing directions. Over millennial timescales ice divides may coalesce, migrate or disappear, so that the directions of ice movement change in response. Ice movement over large areas of ice sheets and ice caps is generally very slow, but corridors of much more rapid movement called *ice streams* evacuate most of the ice mass away from ice domes and ice divides. Towards the margins of ice sheets and ice caps, where the ice becomes thinner, ice movement becomes focused in *outlet glaciers*, which are valley glaciers that are fed from an ice sheet or ice cap.

The most extensive topographically constrained glaciers take the form of *icefields*, which comprise a *transection complex* of glaciers that cover most of the terrain but are separated by mountain summits and ridges that rise above the ice surface as *nunataks* and focus ice movement along intervening cols and valleys. Peripheral tongues of ice draining icefields along valleys are termed *valley glaciers* (Fig. 4.2a). Valley glaciers may also form independently, fed by ice flowing from high-level hollows in mountainous areas called *cirques*, or, in Scotland, *corries*, a term derived from the Gaelic word *coire*, denoting a mountainside hollow or valley head. Where valley glaciers extend outwards from a narrow mountain valley onto adjacent low ground they tend to spread out laterally, forming a broad *piedmont glacier*. Small glaciers restricted to corries are described as *cirque glaciers* or *corrie glaciers* (Fig. 4.2b) and very small glaciers perched high on mountainsides or corrie backwalls are described as *niche glaciers*.

**Figure 4.2** (**a**) Valley glaciers (outlet glaciers) fed by icefields on Axel Heiberg Island, Canadian Arctic (photograph by David Evans). (**b**) Two corrie glaciers in northern Sweden (photograph by John Gordon). Both pictures also show end moraines and lateral moraines deposited when these glaciers were more extensive.

Glaciologists also classify glaciers by the temperature of the ice. *Temperate glaciers* are those in which the ice throughout the glacier is at melting point and is referred to as 'warm' ice; such glaciers typically occur in mid-latitude mountain areas, such as those of New Zealand and western Norway. *Cold glaciers* are those in which all the ice is below melting point ('cold' ice); such glaciers are found in arid polar areas, such as parts of the Antarctic and the Canadian High Arctic. *Polythermal glaciers* are those that contain both warm ice and cold ice, usually with a layer of cold ice overlying warm ice, and occur in a wide range of environments. These distinctions are important, because cold glaciers are frozen to the underlying substrate so that subglacial erosion is usually very limited, whereas temperate glaciers and warm-ice-based polythermal glaciers slide over and erode the underlying terrain. Another distinction is that meltwater is usually confined to the surface and margins of cold ice glaciers, but present within and at the base of warm ice glaciers.

## Accumulation, ablation and mass balance

The size of a glacier reflects two competing influences: the rate of ice accumulation on its higher reaches, and the rate of ice ablation (melting) at lower altitudes. The higher parts of a glacier are termed the *accumulation zone*, because here the rate of snow accumulation and ice formation exceeds the rate of snow and ice melting, so there is a net gain in ice volume. The lower parts of a glacier form the *ablation zone*, where melting of snow and ice exceeds accumulation, so there is a net loss in ice volume. Glacier movement transfers ice from the accumulation zone to the ablation zone, so that although the ablation zone loses ice each year through melting, this is replenished by surplus ice flowing from the accumulation zone (Fig. 4.3), much like transferring money from a bank account that is in credit to one that is in deficit.

The boundary between the accumulation zone and the ablation zone is termed the *equilibrium line*, at which there is neither net gain nor net loss of ice. The altitude of this boundary varies from year to year: after a warm summer it may be high up on a glacier, but after a cool summer it may be at a much lower altitude, so the term *equilibrium line altitude* (ELA) is used to refer to the average position of the boundary over a period of years or decades. If there is longer-term climatic cooling and/or an increase in winter snowfall, net accumulation will exceed net ablation, the ELA shifts to a lower altitude and the glacier terminus advances, increasing the area of

the ablation zone until the overall balance between ice accumulation and ice ablation is restored. Conversely, during a period of climatic warming, ablation exceeds accumulation, the ELA migrates up-glacier and the glacier terminus recedes, shrinking the ablation zone area until a balance between accumulation and ablation is regained. If warming continues, glaciers may disappear entirely, a fate that awaits many small mountain glaciers today. It is important to appreciate that the concept of 'glacier retreat' involves shrinkage of the volume of ice and recession of the lower margin of the ice; there is no change in the downglacier direction of ice movement. The relative values of ice accumulation and ice ablation on a glacier are expressed as its *mass balance*: if net accumulation exceeds net ablation over a period of years, then a glacier has a positive mass balance and advances, but if net ablation exceeds net accumulation, a glacier has a negative mass balance and retreats.

The main source of ice accumulation on glacier surfaces is snowfall, on some glaciers augmented by snow blowing or avalanching from adjacent terrain. Snow accumulating in the ablation zone of a glacier usually melts during the summer, but in the colder, higher accumulation zone the snowpack gradually thickens, and changes first to firn (dense, hard-packed snow) then to ice. In the ablation zone, melting of ice occurs both at the glacier surface, in response to solar radiation and above-freezing air temperatures, and within or below the glacier where the ice is thawed by contact with meltwater streams. Glaciers terminating in the sea or deep lakes also lose mass by *calving*, which involves large blocks of ice toppling from the glacier snout and floating away as icebergs.

## Glacier movement

An interesting property of ice is that it behaves in different ways depending on the amount of stress (pressure)

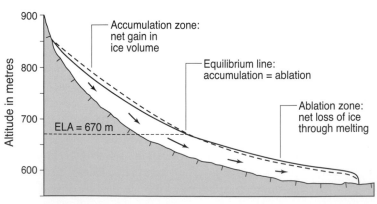

**Figure 4.3** Schematic cross-section through a small mountain glacier, illustrating net ice gain in the accumulation zone, net ice loss in the ablation zone and movement of ice from the former to the latter across the equilibrium line at 670 m. If net accumulation of ice increases, the equilibrium line shifts downglacier and the glacier advances, increasing the area of the ablation zone until a balance between ice accumulation and ice ablation is restored. The opposite happens if net accumulation is reduced or the rate of ablation increases: the equilibrium line moves up-glacier and the ablation zone shrinks, causing the glacier terminus to retreat.

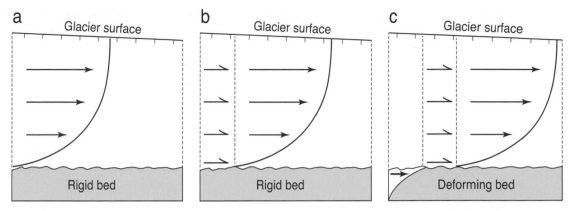

**Figure 4.4** Velocity profiles within hypothetical glaciers. (**a**) Cold-ice glaciers move by internal deformation only, as they are frozen to the underlying substrate. Deformation increases with depth; the upper layers of the glacier are carried passively forward on underlying deforming ice. (**b**) Warm-ice glaciers move by internal deformation and by sliding over (and eroding) the underlying substrate. (**c**) If the glacier bed consists of saturated unfrozen sediment, the top of the sediment may deform, adding to the total velocity of warm-ice glaciers.

that it is under. The ice from a freezer, for example, is a brittle solid, and the same is true of ice near the surface of a glacier, where stresses are low. At greater depths, where the forces induced by the weight of the overlying ice are greater, ice behaves in a different way, 'flowing' or deforming along the direction of the glacier's surface slope. All glaciers experience movement by such *internal deformation*, which increases with depth within the ice (Fig. 4.4). Cold glaciers move by internal deformation alone, but temperate glaciers and warm-ice-based poly-thermal glaciers also move by sliding over the underlying substrate, a process aided by the presence of water at the base of the ice. Moreover, if warm-ice-based glaciers are underlain by water-saturated sediments rather than bedrock, the top layers of sediment may deform in the direction of ice movement, adding to the sum of glacier movement (Fig. 4.4).

Rates of glacier movement depend on the surface slope and mass balance of a glacier, and the temperature of the ice. If a temperate glacier has a steep mass balance gradient, with high rates of both accumulation and ablation, a large volume of ice is transferred each year from the accumulation zone to the ablation zone and ice velocities are high: the surface velocity of the Franz Josef glacier in New Zealand, which has a steep mass balance gradient, is up to 600 metres per year. Most temperate glaciers have maximum surface velocities of a few tens to a few hundreds of metres per year. Polar ice sheets or ice caps with very gentle mass balance gradients move at rates of just a few metres per year, though fast-flowing ice streams within such ice masses may achieve maximum surface velocities of over 1000 metres per year

in locations where the flowing ice has a very large source area. Some glaciers also undergo periodic *glacier surges*, periods of short-term rapid frontal advance with surface velocities of up to about 30 m per day.

## Glacial erosion, transportation and deposition

Erosion by successive ice sheets and valley glaciers has been the dominant process responsible for fashioning the mountain landscapes of Scotland. Glacial erosion of resistant bedrock is rather slow, however, and the major glacial landforms of the Highlands (such as glacial troughs and rock basins) reflect the cumulative effects of glacial erosion over much of the Pleistocene epoch. Two main processes are involved, both associated with warm-ice-based glaciers that slide over their beds.

The first process is *glacial plucking* or *glacial quarrying*. Glacier ice passing over a bedrock obstacle tends to undergo melting on the up-glacier side of the obstacle (where the stress operating at the glacier bed is locally increased, thawing the ice) and refreezing on the downglacier side, where the stress is reduced. If the rock is fractured, water under very high pressure dislodges chunks of rock that become frozen into a thin belt of debris-rich ice at the base of the sliding glacier. The second process is *glacial abrasion*: rock fragments embedded in the base of the sliding glacier scrape over underlying bedrock surfaces, forming long narrow scratches or grooves called *striae* and chiselling away flakes of rock, forming small concentric fractures or *chattermarks* (Figs 1.2 and 4.5). Abrasion releases silt-sized and sand-sized particles and small rock fragments that also became frozen into the base of the ice. The

**Figure 4.5** Striae (parallel grooves) and chattermarks on a glacially smoothed outcrop of Torridonian sandstone. The striae were etched across the bedrock surface by clasts frozen into the base of a former glacier. The chattermarks represent chiselling of rock fragments from the outcrop by a subglacial boulder held in the grip of the sliding ice.

effectiveness of abrasion depends not only on the force exerted by the weight of the overlying ice and the rate of sliding, but also on the resistance of the underlying rock: strong metamorphic or igneous rocks such as quartzite, basalt or gneiss tend to be abraded much more gradually than weakly bonded sedimentary rocks such as sandstones and shales.

As glaciers and ice sheets advance, they also move over terrain covered by earlier sediments. These may consist of glacial deposits emplaced by a previous glaciation, alluvial or deltaic sediments laid down by rivers, or debris deposited by rockfalls and landslides. Over time, these sediments are eroded by the overlying ice, mainly by being frozen into the glacier sole. Corrie and valley glaciers also receive debris from adjacent mountain slopes, transported onto the ice by rockfalls, landslides and snow avalanches. If such debris reaches the glacier surface in the accumulation zone, it becomes buried within the accumulating ice, and may travel *englacially* (within the body of the glacier), re-emerging at the ice surface in the ablation zone. Debris deposited directly onto the ablation zone remains on the ice surface and is carried *supraglacially* towards the glacier terminus. The

transport pathways of stones and boulders deposited by glaciers can be interpreted from their form. Those that have been in *active transport* at the base of the ice are usually edge-rounded, faceted and striated as a result of being scraped over the underlying bedrock. Rock fragments that have experienced *passive transport* on or within the ice retain their original characteristics and are typically angular.

Sediments deposited by glacier ice or by glacial meltwater streams are collectively termed *glacigenic deposits*, though many geomorphologists use the more informal term *drift deposits* as a synonym. Debris deposited directly by a glacier is known as *till*, an archaic Scots term for a stiff clay soil. Most tills comprise *clasts* (pebbles, cobbles and boulders) embedded in a matrix of fine-grained sediment that may be dominated by sand, silt or clay (Fig. 4.6). Many tills contain *erratics*, which are 'foreign' clasts that have been transported by a glacier from their source outcrop and deposited in an area underlain by a different rock type. Till is deposited from glacier ice in a variety of ways. *Lodgement till* is deposited when a debris-rich layer at the base of the ice becomes plastered onto the underlying substrate;

**Figure 4.6** Two till deposits in Glen Rosa, Arran. Below the lower dashed line is a grey lodgement till, deposited under a former glacier. Between the dashed lines is a bed of windblown sand. Above the sand is a brown ablation till, dumped from the former glacier surface. This sequence implies that retreat of the glacier that deposited the lower till was succeeded by deposition of windblown sand over ice-free terrain, then by readvance of the glacier and deposition of the upper till.

*subglacial meltout till* is also deposited under the ice, if the base of a glacier undergoes thaw; *deformation till* is formed when saturated subglacial sediment undergoes deformation under the moving ice; *supraglacial meltout till* or *ablation till* accumulates on the surface of a thinning glacier during retreat, and is subsequently dumped on the underlying ground; and *flow till* is the name given to supraglacial meltout till that has become saturated and flowed off a glacier. Most of these types of till have a distinctive sedimentological signature: lodgement till, for example, is usually strongly compacted, ablation till is comparatively loose and deformation till contains structures indicative of internal dislocation.

In lowland areas of Scotland, tills deposited and reworked by successive Pleistocene ice sheets often reach thicknesses of tens of metres, forming an undulating *till sheet*. In most areas this completely blankets the underlying sedimentary rocks, so that only resistant igneous rocks (lava flows, volcanic plugs and sill rocks) emerge from the till cover. In some areas, the till sheet has been moulded into streamlined *glacial bedforms*, of which the most common are *drumlins*, smooth, oval or elliptical hills, usually 5–50 m high and up to about 800 m long. Numerous drumlins commonly occur in proximity, as in the case of those that underlie much of Glasgow. In the Highlands, however, drumlins are rare, and the dominant subglacial bedforms consist of low, narrow, regularly spaced ridges, usually less than 3 m high and 3 m wide but up to about 100 m long. These bedforms are termed *flutes* or *fluted moraines*, and like drumlins they are aligned in the direction of former ice movement.

Landforms deposited at the margins of glaciers are termed *moraines*. Various processes contribute to the formation of ice-marginal moraines, notably 'bulldozing' by a glacier, which pushes up sediment at its front and margins, 'dumping' of supraglacial debris around the margins of a glacier, and 'thrusting' of debris-rich layers of ice at the glacier margin, resulting in the formation of ridges of till as the glacier thins and retreats. Moraines are also classified by location: *end moraines* or *terminal moraines* are those deposited at the limit of a major advance or readvance of the ice front and can range in size from small ridges of boulders to massive till ridges tens of metres high (Fig. 4.2b). *Recessional moraines* are nested ridges deposited as the ice margin oscillates during overall retreat. *Lateral moraines* form along the margins of valley glaciers; the outermost lateral moraines delimit the maximum extent of a glacier advance or readvance, whereas those inside the glacial limit are recessional moraines.

A particularly common glacial depositional feature in the Scottish Highlands takes the form of areas of *hummocky moraines*, which are apparently chaotic mounds of till, commonly up to about 20 m high, that litter valley floors occupied by the last mountain glaciers (Fig. 4.7a). Some hummocky moraines represent the products of ice stagnation, dumped in a chaotic way around a rapidly melting glacier terminus that had become cut off from its accumulation area. Others represent 'drumlinoid' or

**Figure 4.7** (a) Hummocky moraines in Drumochter Pass, central Grampians. (b) Hummocky moraines forming concentric arcs in Coire Fee above Glen Clova, southeast Grampians. Each arc represents a recessional moraine marking the former position of the margin of a glacier that occupied the corrie about 12,500 years ago (photograph by Charles Warren).

fluted subglacial bedforms. The great majority of Scottish hummocky moraines, however, form nested chains, often arcuate (crescent-shaped), and represent recessional moraines formed where debris deposition at the glacier margin was uneven (Fig.4.7b).

Finally, *medial moraines* are ridges of debris aligned down-ice on the ablation zones of glaciers. They form in various ways, for example at the confluence of two valley glaciers or down-ice from a nunatak that supplies the glacier surface with rockfall debris. The apparent thickness of medial moraines on glacier surfaces is illusory: the debris cover protects the underlying ice from melting, so much of the moraine consists of a core of ice. As a result, most medial moraines have little or no surface expression after deposition, and they are rare in the Scottish landscape.

## Glacifluvial landforms

Melting of glacier ice produces copious amounts of meltwater in the summer, so that rivers draining the ice typically have high flood discharges. *Glacifluvial landforms* are those formed through erosion or sediment deposition by glacier-fed meltwater streams. On cold glaciers, meltwater flow is usually restricted to the glacier surface or margins, as water entering the ice via crevasses tends to freeze on contact with the ice. Meltwater associated with temperate or polythermal glaciers, however, also descends into the ice via crevasses or vertical shafts, then flows within the ice along englacial tunnels or descends to the base of the ice to flow along *submarginal* routeways (along valley sides) or *subglacial* routeways (along valley floors).

The former routeways of glacial meltwater often take the form of *meltwater channels*. Thousands of meltwater channels marking the former routes followed by glacial rivers are present in the Scottish landscape. Most were eroded in till during the thinning and retreat of the last Scottish Ice Sheet, though some bedrock channels have a longer history, having been excavated by meltwater erosion during successive glaciations. In upland areas, numerous meltwater channels originated as ice-marginal or sub-marginal channels that cut obliquely across valley sides, sometimes converging or bifurcating. Others are represented by V-shaped notches or gorges that mark the courses of former glacial streams across cols and spurs. Most meltwater channels no longer contain streams, because there is now no catchment area to feed runoff. Many Scottish rivers, however, follow routeways that formerly carried meltwater runoff away from former ice margins, and in some cases this is manifest in 'underfit' streams that occupy anomalously large channels formed by powerful meltwater rivers.

High-energy *proglacial* meltwater rivers flowing away from a glacier snout often form *braided river systems*, characterized by multiple channels that diverge and converge, and frequent shifts of channel position during floods (Fig. 4.8a). The floodplains of such rivers are variously referred to as *outwash plains*, *braidplains* or by the plural Icelandic term *sandar* (singular *sandur*). These are areas of net sediment accumulation, where meltwater streams deposit copious volumes of sand and gravel, in the form of *outwash deposits* or *glacifluvial deposits*. Unlike most tills, these are often stratified, and clasts are typically rounded by abrasion. When deposited in association with glacier ice, outwash deposits form

**Figure 4.8** (a) Sandur (floodplain crossed by a braided river) in front of Dogleg Glacier, Ellesmere Island, Canadian Arctic Archipelago (photograph by David Evans). (b) Ice-dammed lake impounded by the Perito Moreno Glacier, an outlet glacier of the Patagonian Ice Cap, Argentina (photograph by John Gordon).

a number of distinctive landforms (*kettle holes*, *kames*, *kame terraces* and *eskers*) that are described in the next chapter.

### Glacial lakes

Glacial lakes or *ice-dammed lakes* form where drainage is blocked by a glacier so that water accumulates at the glacier margin. This occurs, for example, where a glacier advances up a valley, or across the mouth of a tributary valley (Fig. 4.8b). Such ice-dammed lakes commonly drain across the lowest available ice-free col via an *overspill channel* that is abandoned when the lake eventually drains. Features diagnostic of the former presence of an ice-dammed lake include abandoned shorelines and deltas, and stratified silty lake-floor deposits known as *varves*, with each varve representing one year's sediment deposition. As the damming glacier thins and retreats, the lake waters may be released as a catastrophic flood known as a *jökulhlaup*.

### Paraglacial landscape modification

Although many of the glacial and glacifluvial landforms and sediment types described above are preserved in a pristine state after deglaciation, the downwastage and retreat of glaciers in mountain areas is often followed by modification of the glacial landscape by paraglacial processes, defined as geomorphological processes 'conditioned' by previous glaciation and deglaciation of the terrain. Mountainsides released from the weight of glacier ice experience failure in the form of rockfalls, rockslides and large-scale rock-slope deformation (Chapter 7); glacial drift on slopes may be reworked by landslides, mountain torrents, debris flows and avalanches; and glacial deposits on valley floors are commonly eroded, transported and redeposited by rivers. The operation of paraglacial processes may be envisaged as a sediment cascade, in which non-renewable, glacially conditioned sediment sources (unstable rockwalls, drift-mantled hillslopes and valley-floor glacigenic deposits) are released or entrained by non-glacial processes and transported to sediment sinks that represent their final resting place, usually in floodplain deposits or in lakes or the sea (Fig. 4.9). However, source-to-sink transport is often interrupted by sediment redeposition in paraglacial sediment stores, notably in rockfall accumulations (talus), debris cones, alluvial fans, river terraces and floodplains. Many postglacial depositional landforms in mountain areas of Scotland are relict features

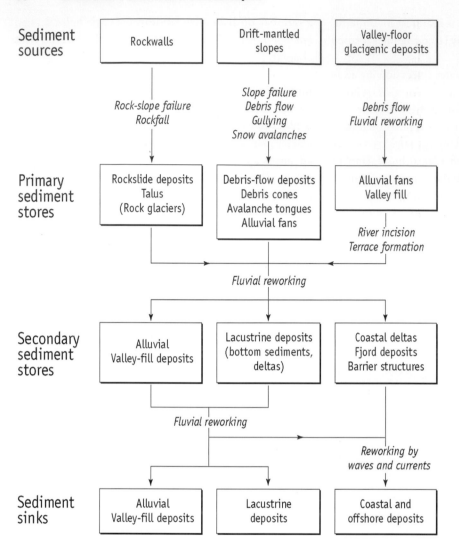

**Sediment sources**

| Rockwalls | Drift-mantled slopes | Valley-floor glacigenic deposits |

*Rock-slope failure*
*Rockfall*

*Slope failure*
*Debris flow*
*Gullying*
*Snow avalanches*

*Debris flow*
*Fluvial reworking*

**Primary sediment stores**

| Rockslide deposits Talus (Rock glaciers) | Debris-flow deposits Debris cones Avalanche tongues Alluvial fans | Alluvial fans Valley fill |

*River incision*
*Terrace formation*

*Fluvial reworking*

**Secondary sediment stores**

| Alluvial Valley-fill deposits | Lacustrine deposits (bottom sediments, deltas) | Coastal deltas Fjord deposits Barrier structures |

*Fluvial reworking*

*Reworking by waves and currents*

**Sediment sinks**

| Alluvial Valley-fill deposits | Lacustrine deposits | Coastal and offshore deposits |

**Figure 4.9** Trajectories of paraglacial sediment transfer. Following deglaciation, sediment derived from unstable rockwalls, drift-mantled slopes and valley-floor glacigenic deposits accumulates in paraglacial *sediment stores* such as talus cones, debris cones, alluvial fans or valley-fill deposits (floodplains). Diminishing sediment supply results in erosion of sediment stores, mainly by rivers, and transport of sediment towards terminal *sediment sinks* that represent its final resting place. Adapted from Ballantyne, C.K. (2002) Paraglacial Geomorphology, *Quaternary Science Reviews* 21, 1935–2017. © 2002 Elsevier Science Ltd.

of paraglacial origin, on which net erosion has now replaced deposition as the dominant mode of geomorphic activity (Chapters 7 and 9).

## The last Scottish Ice Sheet
### Ice-sheet growth and maximum extent

Although Scotland experienced numerous periods of Pleistocene glaciation, our knowledge of glacial events before the expansion of the last Scottish Ice Sheet is limited. This is because the last ice sheet buried or obliterated most evidence of previous glaciations, though at a few locations glacial deposits laid down by earlier ice sheets have been identified in quarries and gravel pits. The last glacial stage extended from ~130 ka to ~11.7 ka and is known in Britain as the *Devensian* and in Europe north of the Alps as the *Weichselian*. The last Scottish Ice Sheet expanded and contracted during the

*Late Devensian* (= *Late Weichselian*) glacial substage, which lasted from ~31 ka to ~11.7 ka.

The last ice sheet grew from an ice cap of limited extent that occupied the western Highlands after ~40 ka. At this time, lowland parts of Scotland were unglaciated but underlain by permafrost (perennially frozen ground) and covered by tundra vegetation indicative of very cold, rather arid conditions. Subsequent expansion of the ice sheet across lowland areas has been established by radiocarbon dates obtained from organic material buried under glacial deposits, including the antlers of reindeer and bone fragments from woolly rhinoceroses that grazed the lowland tundra before it was buried by advance of glaciers from the Highlands. These dates indicate that expansion of the Highland glaciers across lowland and nearshore areas occurred between ~35 ka and ~32 ka.

As the Highland Ice Cap expanded, it met and became confluent with independent ice caps that formed on other upland areas of Scotland, notably on the Southern Uplands, Galloway Hills, Arran, Skye, Mull, the Outer Hebrides, Shetland and the Cairngorms. These independent ice centres fed the growing ice sheet, but also diverted the flow of glaciers emanating from the western Grampians and Northern Highlands. The Skye Ice Cap, for example, diverted mainland ice both northwards into the North Minch, and southwards into the Sea of the Hebrides. Similarly, the Outer Hebrides acted as a major independent centre of ice accumulation, such that only the northernmost tip of Lewis was over-run by ice moving northwest from the mainland. To the south, ice from the southwest Highlands linked with the expanding Southern Uplands Ice Cap to form a major north–south ice divide, from which ice flowed both east towards the North Sea and west across the continental shelf north of Ireland. These various independent ice centres appear to have persisted throughout the lifetime of the last Scottish Ice Sheet. The surface of the ice sheet was therefore not a single broad dome of ice, but a complex of interconnecting ice domes and ice divides that slowly migrated through time, causing shifts in the directions of ice flow.

Moreover, as Figure 4.10 shows, the last Scottish Ice Sheet was just one component (albeit the dominant one) of the last British–Irish Ice Sheet, which covered not only Scotland, but also Ireland, most of Wales, and northern and western England. To the south, the British–Irish Ice Sheet at its maximum extent impinged on the Scilly Isles, reaching latitude 49°N. To the east, ice from Scotland met and became confluent with ice advancing westward from Norway (the Fennoscandian Ice Sheet) in

**Figure 4.10** The maximum extent of the last British–Irish Ice Sheet, which was confluent with the Fennoscandian Ice Sheet in the North Sea Basin. Note that the extent of the ice sheet in some offshore sectors is uncertain, and that the ice sheet did not reach its limit in all sectors simultaneously.

0    100    200 km

the North Sea Basin; this powerful ice sheet diverted ice moving east from Scotland both to the northwest, across Caithness and Orkney, and southwards along the east coast of England, ultimately as far as Norfolk. To the west, the ice sheet reached the edge of the continental shelf along much of its length but terminated short of the shelf edge in some sectors. It failed to over-run the St Kilda archipelago, 40–60 km from the edge of the continental shelf; St Kilda was the only part of the present land area of Scotland that was not covered by the last ice sheet, though it supported its own tiny glaciers.

As the last Scottish Ice Sheet advanced across the adjacent continental shelf, sea level was much lower than now. This is because enormous volumes of water were drawn from the global oceans to feed the expansion of the great northern hemisphere ice sheets. By 26 ka, global sea levels had fallen by about 125 m, so that glacier ice advancing across the continental shelf or North Sea Basin was grounded rather than floating. Westward termination of much of the British–Irish Ice Sheet at the shelf edge was caused by the ice margin entering deep water, where it began to float, calve and release icebergs.

An intriguing feature of the last British–Irish Ice Sheet is that it reached its maximum extent at different times in different sectors, so technically the map depicted in Figure 4.10 is fictitious: at no single time did the ice sheet reach all the ice limits shown. The ice sheet appears to have achieved its maximum westwards and northwestwards extent between 30 ka and 27 ka but reached the Scilly Isles no earlier than 26–25 ka and south Yorkshire possibly as late as 22–21 ka. These contrasts in timing imply that the northwestern part of the ice sheet had begun to retreat even as more southerly sectors were still advancing. This apparently anomalous behaviour probably reflects reduction in the snowfall feeding ice accumulation in the north, possibly due to more extensive sea-ice cover in the northern Atlantic Ocean, which would have reduced the moisture carried by northern Atlantic airmasses and thus the snowfall over the ice.

## Ice streams, binges and purges

As outlined earlier, *ice streams* are corridors of fast-flowing ice within ice sheets. Their velocities can reach several hundreds of metres per year, and their margins are sharply defined by crevassed shear zones. The locations of former ice streams (*palaeo-ice streams*) that drained the last Scottish Ice Sheet have been identified by a range of criteria, notably streamlined and elongate glacial bedforms that have developed in bedrock or sediments, a well-defined flow track and evidence of convergent ice flow. Palaeo-ice streams flowed eastward along the Tweed Valley and Forth Valley to feed a major ice stream that extended down the east coast of England, and another flowed out of the Moray Firth into the North Sea Basin. To the northwest, the Minch palaeo-ice stream was fed by ice flowing from the northwest Highlands, Skye and eastern Lewis, and occupied a submarine trough that extends ~200 km to the shelf edge. Another palaeo-ice stream drained ice from the western Highlands and Hebrides westwards across the Sea of the Hebrides to the shelf edge northwest of Ireland. The largest of all the former ice streams was the Irish Sea Ice Stream, which was fed by ice from southwest Scotland, the English Lake District, Wales and Ireland, ultimately draining at least 17% of the mass of the British–Irish Ice Sheet and terminating in the Celtic Sea at 49°N (Fig. 4.10).

The detailed history of these palaeo-ice streams during the build-up and retreat of the last ice sheet is unknown, but it seems that the ice sheet underwent several 'binge-and-purge' cycles during its existence. During 'binges', cold-based ice thickened over upland ice centres, but then slight changes in climate or glacial dynamics destabilized the thick ice cover, causing 'purges' when the ice sheet drained along ice streams, drawing down the ice cover over upland areas. Thus the last Scottish Ice Sheet did not simply thicken progressively during its overall expansion then thin during its retreat, but oscillated between thicker and thinner phases.

## Did the last ice sheet cover the Scottish mountains?

Until recently, it was widely accepted that the summits of some Scottish mountains remained above the surface of the last ice sheet as nunataks even when the ice sheet was at its thickest. This view was based on geomorphological evidence: the flanks of many mountains exhibit glacially scoured bedrock, but their upper slopes and summits are occupied by *blockfields* (rubble formed by frost action), *tors* (bedrock towers) and outcrops of shattered rock, and exhibit little or no evidence of erosion by glacier ice. The boundary between the lower ice-scoured slopes and the frost rubble on higher slopes forms a *trimline* that marks the uppermost altitude to which glacial erosion has 'trimmed' the cover of frost debris.

However, erratic boulders occur above periglacial trimlines on some mountains. On the summit plateau of Maol Chean-dearg in Torridon, for example, erratics

**Figure 4.11** Erratics of white quartzite resting on a blockfield of pink Torridon sandstone boulders at the summit of Maol Cheann-dearg (933 m) in Wester Ross. TCN exposure dating of these erratics and others on nearby summits shows that they were deposited around 16 ka, demonstrating that the last ice sheet overtopped all summits in Wester Ross. Preservation of the blockfield indicates that the ice sheet must have been cold-ice-based over mountain summits in this area and thus failed to erode the blockfield debris.

of white quartzite are littered across a blockfield of sandstone boulders weathered from the underlying bedrock (Figs 1.3 and 4.11). TCN dating of the exposure age of such high-level erratics has demonstrated that they were deposited by the last ice sheet, implying that all Scottish mountains were buried under glacier ice when the last ice sheet was at its thickest. Numerical models that attempt to simulate the growth of the last ice sheet suggest that the ice divide over the Scottish Highlands reached an altitude of at least 1400 m above present sea level, and possibly as much as 2500 m.

So why did many Scottish mountain summits escape erosion by the last ice sheet? The answer lies in the temperature of the ice. As noted earlier, warm-based glacier ice at melting point slides over its bed and erodes the underlying bedrock, but cold-based ice is frozen to the underlying substrate and moves only by internal deformation. It appears that on high summits and plateaux that support blockfields, the overlying ice remained cold-based throughout the growth and decay of the last ice sheet, and therefore failed to erode summit blockfields. The trimlines evident on some mountains therefore represent the former boundary *within* the last ice sheet between sliding, erosive warm-based ice in valleys and on lower slopes, and passive cold-based ice occupying higher ground.

## Retreat and demise of the last ice sheet

The retreat pattern of the last ice sheet was complex, being driven partly by reduced snowfall and slowly increasing (though oscillating) temperatures, and partly by rising sea levels that hastened the retreat of calving ice margins along submarine troughs on the offshore shelves. Numerous offshore moraine ridges appear on bathymetric maps of the continental shelf west and north of the Scottish mainland. These moraines demonstrate that the retreating ice margin readvanced on several occasions, forming moraines that mark the limit of each readvance. On land, deglaciation was asynchronous, occurring a few millennia earlier in some areas than others, and spanning the overall time interval ~25–14 ka. The earliest parts of Scotland to be deglaciated were northern Lewis and Cape Wrath, which probably emerged from under the retreating ice before ~23 ka. As late as ~17 ka, however, most of Scotland was still ice-covered, apart from parts of the Outer Hebrides, the extreme northwest mainland, Caithness and eastern coastal areas. By ~16 ka the Solway lowlands, Orkney and Shetland were deglaciated, the ice margin lay amongst the Inner Hebrides and along the west coast, nunataks had emerged from under the ice in Wester Ross, the eastern margin of the ice sheet had retreated to the Cairngorms and the sea had invaded the Forth and Tay estuaries. By 15 ka most of the Southern

**Figure 4.12** The Gairloch moraine, part of a chain of moraines that mark the limit of the Wester Ross Readvance in northwest Scotland. This final major readvance of the retreating ice sheet in this area occurred at ~15.3 ka.

Uplands, Midland Valley, Northern Highlands and Eastern Grampians were ice-free and a residual ice cap occupied the western Highlands.

In a few areas, moraines on land delimit the extent of ice-margin readvances that interrupted overall retreat. The best example is a chain of moraines that cross the peninsulas of northwest Scotland from Applecross to Assynt (Fig. 4.12). These have been dated to ~15.3 ka, and mark the limit of the *Wester Ross Readvance*, the last stand of the retreating ice sheet in this area. In the western Grampians a residual ice cap still existed at ~14.7 ka. This date, however, also marks the onset of very rapid warming at the beginning of the temperate interval known as the Lateglacial Interstade (~14.7–12.9 ka; Fig. 4.13). This warming has been quantified by analysis of the climatic implications of chironomid (midge) assemblages present in bogs at various sites in Scotland. Collectively, these analyses indicate a rapid rise in mean July temperatures (probably within a few decades) from 5–6°C to 11–12°C, only slightly cooler than now. It may be some compensation to harassed campers to know that midges, albeit long-dead ones, have some scientific value.

We cannot trace the final demise of the ice sheet in its heartland areas in the western Grampians and Northwest Highlands, as these areas were subsequently reoccupied

by glacier ice under the renewed cold conditions of the Loch Lomond Stade of ~12.9–11.7 ka (Fig. 4.13). It is possible that fragments of the last ice sheet survived the temperate conditions of the Lateglacial Interstade, but if so they must have been restricted to favourable sites such as high plateaux or high, shaded corries.

## The last mountain glaciers: the Loch Lomond Stade

The term *Loch Lomond Stade* (or *Loch Lomond Stadial*) designates the final period of severe cold to affect Scotland, and represents the chronological equivalent of the *Younger Dryas Stade*, a more general term to refer to this cold interval globally. As Figure 4.13 shows, summer temperatures in Scotland plummeted around 12.9 ka. The reason for this is debatable, but one widely accepted theory is that this was due to catastrophic drainage of a huge ice-dammed lake that occupied the location of the Great Lakes in North America. Enormous amounts of freshwater entered the North Atlantic Ocean and appear to have weakened or shut down Atlantic meridional oceanic overturning (AMOC), which draws warm surface waters from the tropics into the North Atlantic Ocean during interglacials. As a result, cold polar

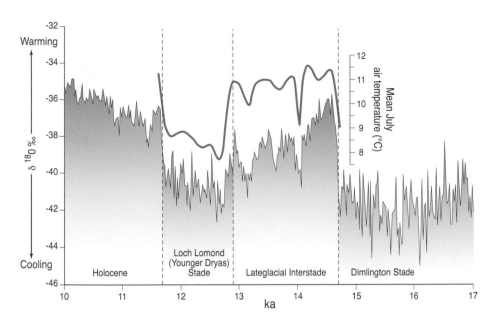

**Figure 4.13** Oxygen isotope fluctuations ($\delta^{18}O$) measured in a core from the Greenland ice sheet record periods of warming and cooling for the period 17 ka to 10 ka. Also shown are mean July temperatures for the period 14.7–11.6 ka, derived from analysis of chironomid assemblages in southern Scotland.

surface water moved southwards, eventually surrounding much of the British Isles with winter sea-ice cover and triggering drastic cooling of northwest Europe. The Loch Lomond Stade ended around 11.7 ka when AMOC was restored, pushing cold polar surface waters back to the latitude of Iceland.

The coldest part of the Loch Lomond Stade probably occurred around 12.5 ka, when mean July sea-level temperatures (reconstructed from chironomids) were no higher than about 8°C in southern Scotland and a degree or two lower farther north. Putting this in context, it implies that summer temperatures 12,500 years ago were similar to November temperatures now. Permafrost was present down to sea level, implying mean annual temperatures of no higher than about -6°C and mean January temperatures in the range -18°C to -22°C. During the Loch Lomond Stade, Scotland therefore experienced a climate fairly similar to the present climate of Spitsbergen (Svalbard) at 78°N latitude.

Glaciers began to grow in corries and on high plateaux in response to climatic deterioration before the onset of the Loch Lomond Stade, then expanded rapidly from their source areas as temperatures fell rapidly around 12.9 ka. This final sortie of glacier ice is known as the *Loch Lomond Readvance*. As these mountain glaciers advanced and coalesced, a major icefield (the West Highland Icefield) formed across much of the western Grampians and Northwest Highlands. When the icefield reached its maximum extent it would have been possible to ski across a continuous cover of glacier ice from Glen Torridon to the southern end of Loch Lomond, or from

Knoydart to Kenmore at the eastern end of Loch Tay (Fig. 4.14). The West Highland Icefield was flanked by satellite ice caps or icefields that developed on peripheral mountain areas, such as Skye, Mull, Assynt, the Monadhliath Mountains and the Drumochter Hills (Fig. 4.15). At the same time, over 100 smaller corrie, plateau and valley glaciers formed on upland areas between the Orkney Islands in the north and the Galloway Hills in southwest Scotland. Despite their relatively high altitude, the Cairngorms also supported mainly small corrie and valley glaciers during the Loch Lomond Stade. This apparent anomaly is explained by the fact that snow-bearing weather systems arriving from the Atlantic had to rise over the West Highland Icefield, which scavenged copious snowfall, leaving northeast Scotland and the Cairngorms in a precipitation shadow with limited snowfall to feed glacier growth; some estimates suggest that the high plateaux of the Cairngorms received less than 600 mm of annual precipitation at this time, and that lowland areas of northeast Scotland may have received less than 300 mm.

As with the last Scottish Ice Sheet, the glaciers of the Loch Lomond Readvance reached their maximum extent at different times. Dating of moraines and related evidence suggests that some achieved their terminal limits as early as 12.8–12.4 ka, whilst others continued to advance until ~12.0 ka. Most glaciers, however, had apparently begun to retreat a few centuries before the onset of rapid climate warming at ~11.7 ka. This may reflect in part a very gradual increase in summer temperatures during the latter part of the stade (12.5–11.7 ka;

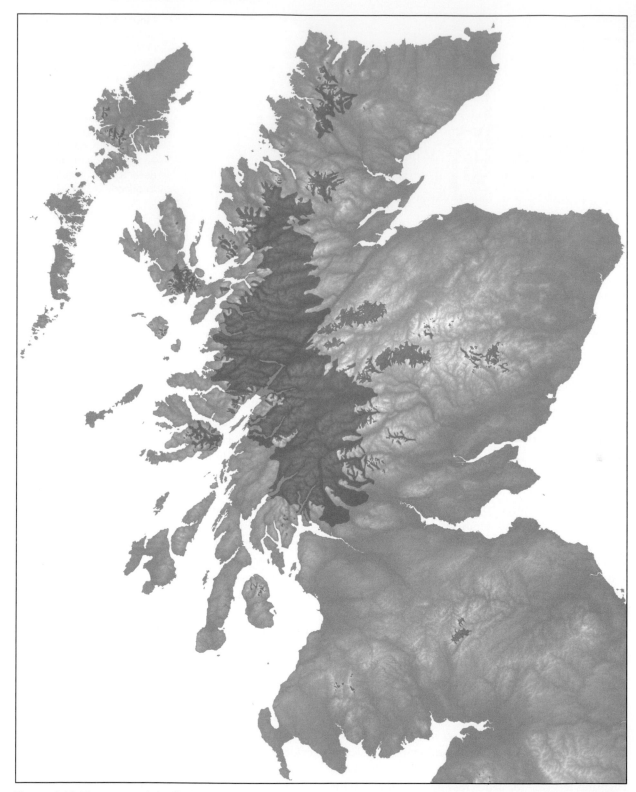

**Figure 4.14** The extent of the final glaciation of Scotland, the Loch Lomond Readvance. Numerous summits protruded through the largest ice mass (the West Highland Icefield) as nunataks, but are not shown here as mapping of the upper limits of the ice in this area is very incomplete. Map by Hannah Bickerdike. Relief derived from the GB SRTM digital elevation model from Sharegeo (www.sharegeo.ac.uk/handle/10672/5), based on a NASA dataset.

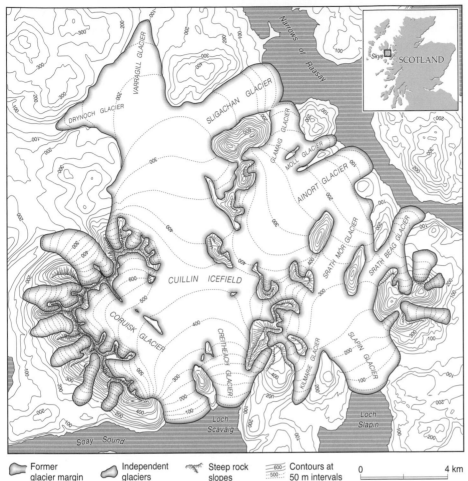

**Figure 4.15** The extent of the Loch Lomond Readvance on south-central Skye, showing a contoured reconstruction of the dimensions of the glaciers. These glaciers reached their maximum extent about 12,500 years ago. Adapted from Ballantyne, C.K. (1989) The Loch Lomond Readvance on the Isle of Skye, Scotland: glacier reconstruction and palaeoclimatic implications. *Journal of Quaternary Science*, 4, 95–108. © Longman Group UK Ltd 1989.

Fig. 4.13), but a progressive increase in sea-ice cover in the North Atlantic may also have reduced the moisture content of westerly winds, causing a reduction in snowfall across Scotland and consequent glacier retreat. It is likely that all glaciers had disappeared from Scotland by ~11.0 ka, by which time summer temperatures approximated those of the present.

In a few areas, such as the Monadhliath Mountains and east Drumochter Hills, plateau ice caps formed on high ground during the Loch Lomond Stade, but elsewhere mountain summits remained above the level of the ice surface and were exposed to the operation of periglacial processes dominated by freezing and thawing of the ground (Chapter 6). Even summits near the thickest part of the West Highland Icefield (near Rannoch Moor) remained as nunataks protruding through the ice surface. In more peripheral areas, the West Highland Icefield fed valley glaciers that radiated outwards into surrounding glens, such as those now occupied by Loch Rannoch, Loch Etive and Loch Tay. Similarly, mountain summits remained above the peripheral icefields of Skye, Mull, Assynt and the West Drumochter Hills (Fig. 4.15). Thus most summits above 600 m in Scotland have remained free of glacier ice since the thinning of the last ice sheet between ~17 ka and ~15 ka, though many of the moraines and glacial deposits in corries, on lower slopes and in valleys were formed much later, during the Loch Lomond Stade.

## The Holocene

The Holocene epoch began at 11.7 ka when resumption or strengthening of Atlantic meridional circulation returned warm oceanic waters to the latitude of Scotland, causing rapid climate warming and disappearance of residual Loch Lomond Readvance glaciers. Temperate conditions similar to those of the present were established by ~11.0 ka, and much of the Holocene prior to ~6.0 ka was probably slightly warmer than now. The second half of the Holocene witnessed a very gradual cooling trend during which 'average' temperatures oscillated by up to 2°C over decades or longer, a trend that has been arrested by recent warming associated with anthropogenic greenhouse gas emissions. In the northern hemisphere as a whole the general pattern of Holocene warming then gradual cooling was interrupted by brief cooling events centred around 9.2, 8.2, 6.3, 4.7, 2.7 and 1.5 ka. The nature of climate change in Scotland at these times is uncertain, but changes in the nature of peat accumulation indicate several shifts in wetness (precipitation) over the past five millennia.

Apart possibly from the 8.2 ka cooling event, the most severe period of Holocene cooling in Scotland was the 'Little Ice Age', a period of general climatic deterioration conventionally attributed to the 16th–19th centuries AD, though cooling may have begun two or three centuries earlier. Historical evidence shows that particular decades during the Little Ice Age were characterized by cool, wet summers, exceptional storminess and survival of perennial snowcover on high ground: summer temperatures in the Cairngorms, for example, averaged about 1°C lower than the 1961–1990 average. Some researchers have suggested that tiny niche glaciers developed in some Cairngorms corries during the Little Ice Age, but the evidence for return of glacier ice at this time is inconclusive.

## Synthesis

Although several prolonged ice ages have occurred during the Earth's long history, *the* Ice Age began only 2.6 million years ago, at the start of the Quaternary period. The Quaternary Ice Age involved numerous shifts from glacial to interglacial conditions and back again in response to changes in the solar radiation budget of northern high latitudes that were amplified by feedback effects. In Scotland, such climate changes have been accelerated by replacement of warm surface water in the North Atlantic by cold polar water (or *vice-versa*) and radical shifts in the trajectory of warm maritime airmasses. In consequence, Scotland has experienced multiple episodes of glaciation during the Quaternary, some involving the growth and decay of icefields and valley glaciers in mountain areas, but others involving the expansion and contraction of thick ice sheets that covered the entire land area and extended far out onto adjacent continental shelves. The last Scottish Ice Sheet began to thicken and expand around 35–32 ka, eventually burying all mountain summits, extending far across the continental shelf to the west, meeting ice advancing from Norway in the North Sea Basin and feeding ice streams that extended south of the Scilly Isles and into Norfolk. Retreat of the last ice sheet was asynchronous but much of the Scottish mainland was still ice-covered at 17 ka. By 15 ka, however, most of the Southern Uplands, Midland Valley, northwest Scotland and the Eastern Grampians were deglaciated, though a residual ice cap occupied the western Highlands.

There followed a brief (~14.7–12.9 ka) period of warmer climate, the Lateglacial Interstade, when glaciers either disappeared completely or were confined to high plateaux or corries. Renewed climatic deterioration before ~12.9 ka caused regeneration and advance of mountain glaciers, which reached their maximum extent during the Loch Lomond Stade of ~12.9–11.7 ka. At this time the West Highland Icefield reoccupied much of the western Grampians and Northwest Highlands, peripheral icefields or ice caps formed on Skye and Mull, in Assynt and on the Monadhliath Mountains and Drumochter Hills, and numerous independent corrie and valley glaciers formed in mountains from Hoy in the north to the Galloway Hills in the south. All mountain glaciers had probably disappeared by ~11.0 ka, and there is no conclusive evidence for the reappearance of glaciers in Scotland under the much milder conditions of the Holocene.

The glacial landforms recording these events are described in the following chapter. Some of these, such as glacial troughs and corries, were formed during multiple glacial–interglacial stages, and are ancient landforms that began to develop during the early Quaternary over a million years ago. Others, such as end moraines and hummocky moraines, are much younger landforms deposited by the last mountain glaciers, some as recently as 12,000 years ago.

# Chapter 5

# Glacial landforms

## Introduction

The year 1840 marked the beginning of our understanding of the dominant role that glacier ice has played in the evolution of Scotland's scenery. On October 7th of that year *The Scotsman* newspaper published a brief letter from Louis Agassiz, a professor at the University of Neuchâtel and a pioneering glaciologist. Under the byline 'Discovery of the former existence of glaciers in Scotland', Agassiz noted that he had seen near Ben Nevis '…the most distinct *morains* and polished rocky surfaces', concluding that 'the existence of glaciers in Scotland can no longer be doubted' and adding, somewhat patronizingly, 'it appeared to me that you would be glad to be able to announce, in your extensively-read journal, the intelligence of the discovery of so important a geological fact'. Rarely has *The Scotsman* enjoyed such a scoop.

The difficult birth of what became known as the Glacial Theory has passed into legend. Accompanied by William Buckland, the President of the Geological Society, Agassiz had travelled north by stagecoach from Glasgow to Fort William, noting on the way landforms similar to those associated with glaciers in the Swiss Alps. Their journey culminated in Glen Roy, where three horizontal shorelines or 'parallel roads' cut across both valley sides. Dismissing Charles Darwin's interpretation of these features as the uplifted products of marine erosion, Agassiz pronounced them to be shorelines formed at the margin of an ancient ice-dammed lake, an interpretation that requires the former existence of a glacier hundreds of metres thick in neighbouring Glen Spean. Whilst Agassiz travelled onwards to Ireland, Buckland visited Charles Lyell, the most famous geologist of the day, at Lyell's estate at Kinnordy in Angus. Lyell was quickly converted to Agassiz' interpretation, and together with Buckland conducted an energetic hunt for further evidence for former glaciation in Scotland.

In November and December 1840, Agassiz, Buckland and Lyell presented papers on their findings to meetings of the Geological Society in London. The grim, hirsute members of the society were not impressed: the distinguished geologist Roderick Murchison appealed for the Glacial Theory to be denounced, and the debate lapsed into acrimony. But the seed had been sown: published just 25 years later, Archibald Geikie's classic book *The Scenery of Scotland* (dedicated, ironically, to Murchison) was largely devoted to glaciation and glacial landforms. Nowadays the Glacial Theory is uncontested (except possibly by diehard creationists), not only for Scotland, but across the globe. Books devoted to the Ice Age and its associated landforms fill the shelves of university libraries. And all this can be traced back to two men standing at the viewpoint in Glen Roy, and a brief paragraph in a Scottish newspaper.

The glacial landforms of Scotland fall into two categories in terms of their age. Large-scale erosional landforms, such as glacial troughs, rock basins, glacial breaches and corries are the product of erosion by successive glaciers and ice sheets throughout much of the Pleistocene. Depositional landforms such as moraines and small-scale erosional features, however, are no older than the last ice-sheet glaciation (~35–14 ka) or the subsequent Loch Lomond Readvance. Below we consider first the nature of glacial erosional and depositional landforms, then the landforms associated with glacial meltwater rivers and former glacier-dammed lakes.

## Landforms of glacial erosion
### Glacial troughs

The most spectacular manifestations of long-term glacial erosion are *glacial troughs* carved by ice flowing along valleys through mountainous terrain. Textbooks often refer to these as 'U-shaped valleys' (as distinct from 'V-shaped valleys' incised by rivers), but the cross-sectional form of glacial troughs tends to be closer to a parabola (Fig. 5.1), and many are asymmetrical, with a

**Figure 5.1** Glacial troughs. (**a**) Glen Sannox, a trough formed in granite, Isle of Arran. (**b**) The Coire Mhic Nobuil trough, excavated in Torridonian sandstone.

steep rockwall on one side and gentler slopes on the other. Glacial troughs represent deepening and widening of pre-existing valleys through erosion by successive generations of valley glaciers, and by fast-flowing ice within former ice sheets. Theoretical modelling of trough evolution suggests that glacial erosion of V-shaped fluvial valleys was initially focused along the lower parts of hillslopes, gradually widening valleys and steepening slopes until a parabolic cross-profile developed. In reality, however, troughs have formed over multiple glacial–interglacial cycles, so that periods of erosion by valley glaciers and ice sheets have alternated with intervals when they are ice-free. During periods of deglaciation, unloading of valley floors from under the weight of the ice causes slight dilation (*stress release*) of the rocks underlying trough floors and walls. Stress release causes fractures

called *dilation joints* (or *sheeting joints*) to form parallel to the floors and sides of troughs. The formation of dilation joints promotes interglacial rockfalls and rockslides, with the debris from such events being removed during the succeeding glacial episode. Trough evolution therefore results not only from prolonged glacial erosion, but also from valley widening by landslides and rockfall during interglacial periods (Fig. 5.2).

Geomorphologists have classified glacial troughs into three types. *Alpine troughs* are those where the glacier occupying the trough was fed by ice from tributary valleys and corries, exemplified in the Highlands by Glens Lyon, Affric and Strathfarrar. *Icelandic troughs* are those where outlet glaciers were fed mainly from a plateau ice cap, and are characterized by an upvalley *trough head*, where the trough terminates abruptly at a high, steep bedrock step. Glen Clova, Glen Muick and Gleann Einich are good examples of this type. *Composite troughs* are those that are open at both ends. The troughs that radiate from Rannoch Moor, such as Glen Coe, Glen Etive and the Loch Treig gap are composite troughs, as are other *glacial breaches* where glacier ice has eroded the former watershed to form a through valley.

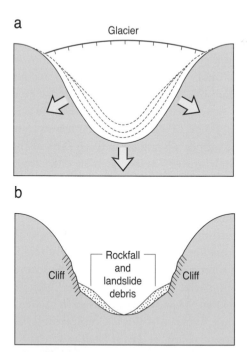

**Figure 5.2** Valley cross-profiles illustrating the evolution of a glacial trough. (**a**) Repeated erosion by valley glaciers and fast-moving ice within ice sheets broadens and deepens the original preglacial valley, with progressive steepening of trough walls. (**b**) During intervening interglacial stages, rockfall and cliff collapse widen troughs. The debris is then evacuated during the succeeding glaciation.

**Figure 5.3** Loch Ericht in the central Grampians. The loch occupies a rock basin up to 150 m deep excavated by successive glaciers along the Ericht–Laidon Fault.

Because troughs have formed in the preglacial valleys cut by rivers, many follow a broadly sinuous course. Straight troughs or sections of troughs frequently follow faults, as is the case with the Great Glen, Loch Maree and Loch Ericht troughs (Fig. 5.3) and the Moffat Valley in the Southern Uplands. In many troughs, glacial erosion has created *truncated spurs*, where spurs aligned athwart the trough axis have been amputated by glacial erosion: the north faces of the Three Sisters in Glen Coe are prime examples. *Hanging valleys* have formed where trough deepening by glacial erosion exceeds that of tributary valleys that are left perched above the trough floor. The floor of the Lost Valley (Coire Gabhail), for example, drops steeply to meet the Glen Coe trough, and spectacular waterfalls mark the junction between hanging valleys and adjacent troughs at An Steall in Glen Nevis and at the Grey Mare's Tail in the Moffat Valley. At many such sites the difference in altitude between the lip of the hanging valley and the trough floor indicates over 200 m of trough deepening by glacial erosion. The deepest troughs with the steepest sidewalls and most regular profiles tend to have formed in relatively homogenous igneous rocks, particularly granite, or in Torridonian sandstones (Fig. 5.1); the schistose rocks that underlie much of the

Highlands tend to be associated with wider troughs with gentler sidewalls. Well-developed troughs are rare in the Southern Uplands, except in the Tweedsmuir Hills and Galloway Hills, both of which were occupied by valley glaciers during periods of limited glaciation and formed major centres of ice dispersal under successive ice sheets.

### Glacial breaches

Glacial breaches occur where ice has crossed a preglacial watershed, eroding a gap. These form two types: *trough breaches*, where erosion by glacier ice has joined troughs on opposite sides of a watershed to form a through valley, and *col breaches*, where ice has crossed the mountain ridge separating one trough from its neighbour, deepening the gaps between summits.

Trough breaches have mainly developed under successive ice sheets, at times when the ice divide (the highest point on the ice sheet) lay some distance from the main watershed. Because ice flows away from ice divides, it is capable of crossing and eroding the terrain separating two troughs, eventually linking them to form a through valley (Fig. 5.4). Trough breaches are common along the spine of the Northern Highlands, where breaching has created about 25 major gaps in

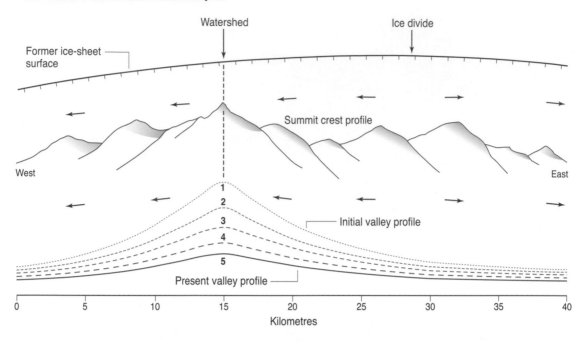

**Figure 5.4** Formation of a trough breach. The ice divide is shown east of the main north–south watershed; arrows represent directions of ice movement. Under successive ice sheets, glacier ice flows westward up-gradient across the watershed separating two troughs. This watershed is progressively lowered by glacial erosion (stages 1–5) until a continuous through-valley (trough breach) has developed.

the main preglacial north–south watershed between Sutherland and Mull, most of which lie over 600 m below adjacent summits (Fig. 5.5). Several of these trough breaches, such as Glen Shiel and Glen Carron, form the corridors for roads connecting the west coast with the rest of Scotland. All appear to have been formed during successive periods of ice-sheet glaciation when the ice divide lay tens of kilometres east of the main north–south watershed, driving ice movement westwards (and against the underlying topographic gradient) up the heads of the long valleys that drain eastward across the Northern Highlands. Similarly, from an ice dome centred on Rannoch Moor, westward-flowing ice created the Glen Coe and Glen Etive breaches in the main north–south watershed. Other apparent breaches, such as those that cut north–south through the Cairngorms (the Lairig Ghru, Lairig an Laoigh and Glen Builg) are less easily explained, as these are aligned athwart the eastward flow of glacier ice that formerly encircled the Cairngorms at the Last Glacial Maximum. These apparent anomalies suggest that at some time in the past an ice divide lay south of the present Cairngorm watershed, forcing ice northwards across cols.

Col breaches, often denoted by the Gaelic word *Bealach*, are widespread throughout the Highlands,

though rare on high tablelands such as the Monadhliath and Gaick plateaus. Most occur above 400–500 m but lie over 200 m below adjacent summits, interrupting the continuity of mountain ridges and posing a challenge to ridgewalkers. Typical examples include Bealach Coire Mhàlagain (695 m), which separates The Saddle (1010 m) from Sgùrr na Sgine (946 m) above Glen Shiel, and the Lochan na Lairig breach (505 m) that links the Glen Lyon and Loch Tay troughs in the Grampians. Col breaches were probably initiated through ice spilling over from one trough into its neighbour during periods of icefield glaciation, when only the higher summits stood above the ice as nunataks, though it is likely that they were deepened by ice movement across cols under successive ice sheets.

## Rock basins and fjords

Nothing illustrates the remarkable erosive capacity of glaciers more than their ability to gouge out deep bedrock basins in the floors of troughs, a feat that no other surface process can emulate. The Highland troughs contain about 70 glacially over-deepened rock basins over 40 m deep, now occupied by long, narrow inland lochs sometimes referred to as *ribbon lakes*. Some of these are remarkably deep: Loch Morar plummets to a depth of 310 m, Loch Ness to 230 m and Loch Lomond

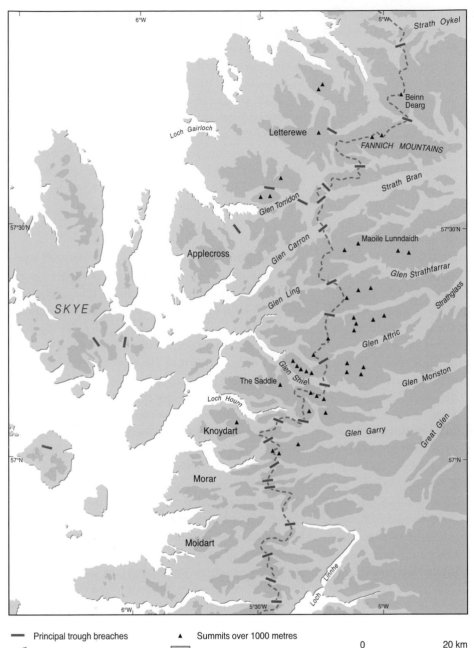

**Figure 5.5** Major trough breaches in part of the Northern Highlands. The majority cut through the main north–south watershed as a result of the ice divide migrating east of the watershed during successive episodes of ice-sheet glaciation.

— Principal trough breaches  ▲ Summits over 1000 metres

/‑‑‑ Main north-south watershed  ▨ Land > 300 metres above sea level

0   20 km

to 190 m, and their capacity is stupendous: Loch Ness contains 7.4 km³ of water, Loch Lomond 2.6 km³ and Loch Morar 2.3 km³.

The location and formation of trough-floor rock basins reflects both geology and topography. Some of the largest basins have been excavated along faults through glacial erosion of the underlying shattered rocks. The Loch Maree and Loch Shin basins follow major faults, the Loch Ericht basin (Fig. 5.3) forms a trench along the Ericht–Laidon Fault, and the basins occupied by Lochs

Lochy, Oich and Ness have been excavated along the Great Glen Fault. Other rock basins have formed where glacier ice radiating from major centres of ice dispersal was confined between steep rockwalls, causing the ice to accelerate and erode the underlying rock, forming basins such as that occupied by Loch Treig, a major outlet for ice draining the Rannoch Moor ice centre. The majority of trough-floor rock basins, however, have formed where confluence of ice from corries and tributary valleys fed accelerated ice flow along an arterial trough. The Loch

Tay basin is a classic example, formed where powerful glaciers emanating from Glen Lochay and Glen Dochart and augmented by ice flowing from adjacent glens and corries gouged out a basin 24 km long and up to 155 m deep. Most of the rock basins that occupy west–east aligned troughs along or east of the main north–south watershed in the Northern Highlands (such as Lochs Fannich, Monar, Mullardoch, Affric, Cluanie and Quoich) were also excavated by accelerated ice flow along trough floors due to convergence of tributary glaciers, both during periods of valley glaciation and within former ice sheets.

The numerous sea lochs (*fjords*) that indent the western seaboard of Scotland from Cape Wrath to Kintyre represent glacial troughs that were flooded as global sea levels rose due to melting of the great continental ice sheets in the closing millennia of the last glacial stage. Like the troughs on land, the location of fjords follows that of preglacial valleys, so they often exhibit a sinuous planform that reflects that of the original valleys (e.g. Loch Sunart, Loch Hourn and Loch Nevis), though fjords or parts of fjords that have developed along faults (such as Loch Linnhe and Loch Shiel) or along the topographic grain (Loch Sween) are straighter. Fjords represent corridors of accelerated ice movement onto the adjacent shelf, and most contain overdeepened rock basins. The floor of Loch Etive, for example, descends to over 130 m below sea level, but the loch shallows at its outlet, forming the powerful tidal overfall known as the Falls of Lora, a foaming maelstrom beloved of kayakers searching for near-death experiences. Many other mainland fjords also exhibit shallowing near their outlets, where glacier flow was less constricted and consequently less erosive. As a result, several fjords have skerries (low rock outcrops, sometimes drowned at high tide) or small rocky ice-moulded islands near their outlets to the open sea. The Summer Isles offshore from Achiltibuie, for example, reflect shallowing of the outlet from Loch Broom, where westward-flowing glacier ice escaped the constriction of the surrounding land.

Studies of sea-floor morphology as revealed by bathymetric surveys have shown that during periods of ice-sheet glaciation the fjords of western Scotland formed the onset zones of powerful, fast-moving ice streams that evacuated much of the mass of the last ice sheet across the continental shelf. Fjords south of Skye fed ice to the Hebrides Ice Stream, which at the Last Glacial Maximum (LGM) formed a corridor of fast-flowing ice that extended to the edge of the continental shelf,

a distance of up to 275 km. North of Skye, the fjords of northwest Scotland fed ice into the Minch Ice Stream, which followed a ~200 km long and 40–50 km wide submarine trough that terminates at the shelf edge. The most remarkable offshore ice stream that formed during the LGM was the Irish Sea Ice Stream. Ice flowing from southwest Scotland was joined by glaciers from England, Wales and Ireland to flow southwards through the Irish Sea to south of the Scilly Isles, implying that ice sourced in Scotland travelled over 800 km to the ice-stream terminus; this single ice stream evacuated 17–25% of all the ice within the last British–Irish Ice Sheet.

## Corries

Corries (usually called *cirques* outside of Scotland) represent some of the most iconic landforms formed by glacial erosion of mountains over multiple glacial–interglacial cycles. They are large bedrock hollows on mountainsides, with a steep headwall and gently sloping floor, and often have a roughly arcuate planform. The floors of some corries consist of a shallow bedrock basin containing a lochan (Fig. 5.6), though many Scottish corries lack this feature; the floors of most consist of glacially abraded bedrock, sometimes buried under moraines, till deposits or peat. Where corries have formed in uniform bedrock, such as those scalloped into granite hillslopes at the margins of the Cairngorms, they are roughly semicircular in planform, with cliffed sidewalls enclosing the corrie floor. Other corries have a less regular form, sometimes reflecting the structure of the surrounding rocks. Some are asymmetrical, with only one sidewall, some lack distinct headwalls (particularly where glacier ice has breached a col), and some have moderately steep floors. Others intersect, producing compound corries, such as the northern corries of the Cairngorms (Coire an Lochain and Coire an t-Sneachda). Corries also occur at the heads of some glacial troughs, such as Coire Odhar at the southern end of Gleann Einich.

Almost all corries in the Highlands and Hebrides have the Gaelic place-name *Coire*, but this name is also frequently applied to valley heads or even deep landslide scars, so many locations labelled *Coire* are not true corries. Archetypal corries with cliffed headwalls, gentle or overdeepened floors and an arcuate planform are easily identified, but mountainside recesses lacking one or more of these characteristics may be ambiguous. A study based only on 'classic' corries recorded 373 examples in all of Scotland, but later research identified 260 corries in the Kintail–Affric area alone, suggesting that

**Figure 5.6** Scottish corries. (**a**) Coire nan Miseach in the Mamores. (**b**) Coire Tuill Bhearnach above Loch Mullardoch. (**c**) Coir'a'Ghrunnda, Cuillin Hills. (**d**) Incipient corries on the northern slopes of Meall a'Bhealaich, Kintail. The hollow on the left is a fluvially incised bedrock gully; that on the right is a landslide scar.

the total for Scotland may amount to over a thousand. This discrepancy results from differences in the criteria used to define what constitutes a corrie.

Most Scottish corries are 200–1200 m long (from headwall to corrie lip) and 200–1200 m wide. Compound corries and trough-head corries are generally larger, with lengths or widths of up to 3 km and occasionally more. There is no clear relationship with rock type: corries occur on almost all rock types in the Highlands and Hebrides. They are rarer in the Southern Uplands, with a handful on the granite of the Galloway Hills and another small group in the Tweedsmuir Hills. A few even occur in the Midland Valley, notably Corrie of Balglass, which has been excavated in the lava scarp that forms the northern flank of the Campsie Fells.

The majority of Scottish corries occupy north-facing or east-facing slopes, and one study found that 86% face between northwest and southeast. This concentration is beautifully illustrated by the mountain ridges

that flank Glen Shiel in Kintail. South of the glen, the northern flanks of the Cluanie Ridge boast 16 corries, whereas north of the glen the sweeping southern slopes of the Five Sisters ridge support just one corrie and two dubious candidates. There are, nevertheless, a few spectacular corries that face south or west, such as south-facing Corrie Brandy and its neighbours above Glen Clova, and the arc of magnificent corries indented into the western flank of the Cuillin Hills on Skye (Fig. 5.6c).

The preferred northerly or easterly orientation of most corries has been explained by persistence of snow-cover on northern and eastern slopes during the early stages of glaciation. Gusty westerly to southerly winds are thought to have blown snow from summits and plateaux so that it accumulated mainly on lee (north- and east-facing) slopes, favouring the eventual formation of small glaciers, particularly at sites where shading reduced summer snowmelt. Most of these embryonic glaciers probably formed in valley heads along the margins of ridges and plateaux or in the hollows formed by deep landslides (Fig. 5.6d). Once glaciers had formed, the initial hollows were enlarged by subglacial erosion, so that during later glacial stages small glaciers reoccupied the same sites, further deepening the hollows. This process is likely to have recurred many times during successive glacial stages.

Studies of the morphology of Scottish corries have demonstrated a correlation between their length, width and depth, suggesting that corrie deepening by subglacial erosion was accompanied by corrie widening both headwards and laterally. Such corrie widening is usually explained by the effects of frost action in causing rockfall and rockslides from the cliffs that form the headwall and sidewalls of corries, and removal of the resultant debris by corrie glaciers. This was probably true when small glaciers occupied the corries and the surrounding cliffs were exposed to frost action, but cannot have been the case during periods of ice-sheet glaciation, when the ice cover thickened to bury mountain summits. Many corries now contain accumulations of rockfall and rockslide debris (Fig. 5.6) that indicate retreat of headwalls and sidewalls after the last glaciers disappeared. Such debris accumulations suggest that corrie widening occurred during successive interglacial periods, particularly due to rockfalls or rockwall collapse caused by unloading of cliffs from under the weight of shrinking ice sheets (Fig. 5.7). Analysis of the morphology of Scottish corries of different size suggests that widening and extension of corries occurred more rapidly than deepening of corrie floors, indicating that the present form of corries owes as much to rockfall activity and rockwall collapse as it does to subglacial erosion.

Another interesting characteristic of Scottish corries is that the altitudes of corrie floors show a general eastward increase, from about 100–400 m in the west (for example on the Outer Hebrides and Skye) to 500–900 m in the Cairngorms and Eastern Grampians. This trend

has been related to dominantly westerly snow-bearing winds and a consequent eastwards reduction in snowfall during the early stages of glacier accumulation. Thicker snow accumulation in the west would have permitted snow to persist through the summer at lower altitudes, eventually forming low-level corrie glaciers. On the

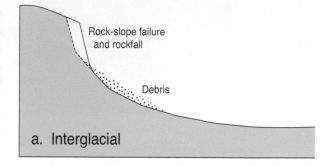

a.  Interglacial

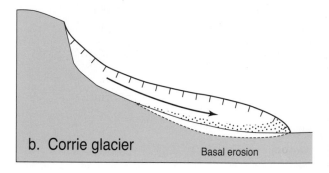

b.  Corrie glacier

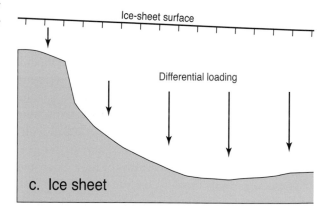

c.  Ice sheet

**Figure 5.7** The cycle of corrie enlargement by rockfall and glacial erosion. (**a**) During interglacials, rockfalls and rockslides cause backwall and sidewall retreat and accumulation of debris. (**b**) In the early stages of glaciation, a corrie glacier removes the rockfall debris and deepens the corrie floor. (**c**) If an ice sheet subsequently forms over the site, basal erosion continues and the weight of the ice causes differential loading. (**d**) After deglaciation, stress release due to unloading results in further rockfalls, rockslides and debris accumulation. Successive iterations of this cycle cause both lateral extension and deepening of the corrie.

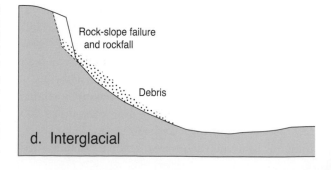

d.  Interglacial

snow-starved eastern mountains, however, thin snow-cover at low altitudes is likely to have melted completely during the summer months, and persisted to form glaciers only at higher, colder elevations. This was certainly the case during the Loch Lomond Stade, when glaciers descended to sea level in the Western Highlands but were confined to ground above about 600 m in the Cairngorms (Fig 4.15).

## Ice-moulded bedrock and landscapes of glacial scouring

As outlined in Chapter 4, glacial erosion of bedrock involves two processes. The first is *glacial abrasion*, which occurs where rock fragments at the base of a sliding, warm-based glacier scour and chisel the underlying bedrock, producing striae and chattermarks and smoothing the rock surface (Fig. 4.5); the second is *glacial plucking* or *glacial quarrying*, where chunks of rock have been excavated from the down-ice end of bedrock obstacles to become entrained in the base of the ice.

On cols, corrie floors, trough floors, plateaux and low-lying terrain underlain by resistant rocks, glacial abrasion under successive ice sheets has produced a range of ice-moulded glacial bedforms. In parts of the western Highlands where abrasion has been intense, these take the form of gently undulating, glacially moulded bedrock that extends over wide areas and is generally aligned along the direction of former ice movement (Fig. 5.8a). Individual streamlined rock knobs rising above such glacially scoured terrain are known as *whalebacks* (Fig. 5.8b), usually a few metres high and a few tens of metres long, and on resistant rocks these sometimes preserve striae (Fig. 5.8c). Other bedrock knobs, known as *roches moutonnées*, are abraded, striated and smoothed on their up-glacier sides, but exhibit glacially-plucked stepped or roughened surfaces on the down-glacier sides (Fig. 5.8d). Such landforms indicate not only the direction of former glacier movement, but also that the ice cover was warm-based and thus capable of sliding over and eroding the underlying bedrock.

In parts of the western Highlands, ice-moulded bedrock dominates the landscape. Much of the mountainous terrain of the peninsulas between Loch Duich and the Sound of Mull contains extensive areas of ice-scoured bedrock that locally extends to mountain summits such as Luinne Beinn (939 m) in Knoydart and Sgùrr Dhomhnuill (888 m) in Ardgour. Some of the most distinctive areas of regional ice scouring, however, occur on low ground or plateaux where glacial erosion

**Figure 5.8** Landforms of glacial abrasion. (**a**) Ice-moulded schist on high ground, Ardgour. (**b**) Whaleback of gneiss, South Harris. (**c**) Gabbro whalebacks near the lip of Coire Làgan in the Cuillin Hills, Skye. (**d**) A roche moutonnée developed on quartzite, Jura; ice movement was from left to right.

**Figure 5.12** An end moraine and paired lateral moraines define the extent of a small glacier that emerged from a corrie on Beinn Dearg Mór (908 m) in Wester Ross during the Loch Lomond Readvance. The lateral moraine on the right is also a drift limit: till blankets the ground inside the moraine but is absent outside the moraine (photograph by Martin Kirkbride).

**Figure 5.13** Bouldery moraines that mark the limit of the Loch Lomond Readvance in Coire Lochain a'Chnapaich in the Beinn Dearg massif, Northern Highlands. The mountain in the background is Meall nan Ceapraichean (977 m).

**Figure 5.14** Part of the Loch Tay lateral moraine on the lower slopes of Ben Lawers. The moraine also defines a drift limit between thin, smooth soliflucted till (deposited by the last ice sheet) on the right and thicker undulating till (deposited by the Loch Lomond Readvance glacier) on the left.

they ascend the eastern side of the Loch Lomond trough to 300 m altitude near Rowardennan. Similarly, the large glacier that advanced northwards through the Treig gap into Glen Spean deposited at its limit an almost continuous moraine, 15 km long, that descends eastwards from 500 m to 400 m across the northern slopes of Chno Dearg, loops around Glen Spean in the form of three massive ridges up to 20 m high, and can be traced westwards along the northern slopes of the glen to an altitude of 400 m. Another remarkable system of lateral moraines snakes intermittently for 25 km around the hillslopes bordering Loch Tay (Fig. 5.14), descending eastwards from 600 m altitude across the southern slopes of Ben Lawers to an end moraine at Kenmore, then ascending the hillslopes south of the loch to terminate at a similar altitude.

In other corries and glens, the limits of Loch Lomond Readvance glaciers lack end or lateral moraines but are represented by *drift limits* or *boulder limits*. Drift limits take the form of an abrupt thickening of till cover, often represented by the appearance of hummocky recessional moraines. As you travel northwards along the A9 road between Calvine and Dalwhinnie, for example, you will notice a transition from smooth hillslopes mantled by soliflucted till to ridges of hummocky till. The hummocky ridges extend northwards across Drumochter Pass to terminate about 4 km south of Dalwhinnie (Fig. 5.15), and represent the area occupied by a Loch Lomond Readvance glacier that was fed by ice from the mountains to the west and splayed out to feed tongues of ice that advanced southwards down Glen Garry and northwards into upper Glen Truim. Other drift limits are marked simply by the junction between till-covered terrain and bedrock outcrops (Fig. 5.12). Boulder limits occur where there is an abrupt termination of spreads of large boulders deposited by the last glaciers; these may terminate in a bouldery ramp or ridge (Fig. 5.13), though this is sometimes absent. The arcuate spread of granite boulders at the head of Glen Rosa in Arran, for example, is a boulder limit that marks the extent of a small glacier that formed during the Loch Lomond Stade.

As they gradually retreated from their maximum extent under the slowly warming summer temperatures and reduced snowfall of the last few centuries of the Loch Lomond Stade, the margins of many glaciers

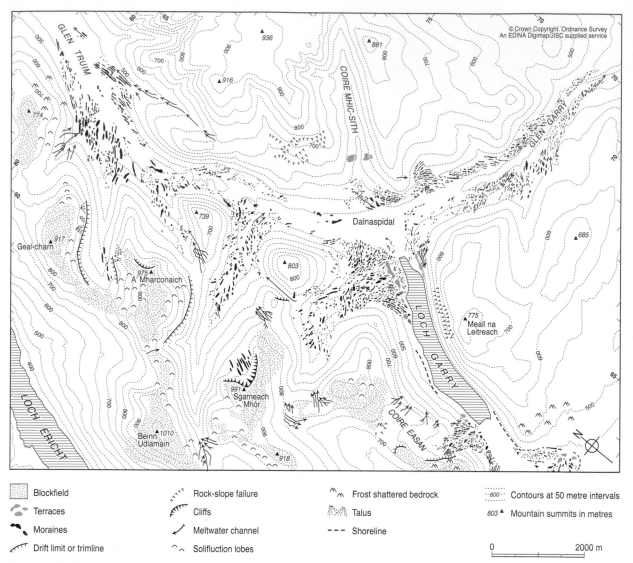

**Figure 5.15** Geomorphological map of the Drumochter Pass area, showing the multiple recessional moraines that mark short-lived readvances of the glacier margins in Glen Truim and Glen Garry during overall retreat of these glaciers from their Loch Lomond Readvance limits. Adapted from Benn, D.I. and Ballantyne, C.K. (2005) Palaeoclimatic reconstruction from Loch Lomond Readvance glaciers in the West Drumochter Hills, Scotland. *Journal of Quaternary Science*, 20, 577–592. © 2005 John Wiley and Sons Ltd.

oscillated, so that numerous brief readvances of the ice margin interrupted overall retreat. In many glens such readvances are recorded by *recessional moraines*, each of which was formed by bulldozing or dumping of sediment at the ice margin, creating nested chevrons of ridges. In many places deposition along former ice margins was uneven, producing *hummocky recessional moraines*. From the valley floor, these often appear as a chaotic assemblage of rounded or conical hummocks, typically 4–20 m high, but viewed from above they resolve into nested ridges or chains of hummocks, each of which records the temporary position of the former

glacier terminus (Fig. 5.16). Recessional moraines, hummocky or otherwise, are common in upland valleys that were glaciated during the Loch Lomond Stade, from the glens of North Harris to the Talla Valley in the Tweedsmuir Hills. By contrast, streamlined subglacial bedforms are fairly rare inside the limits of the Loch Lomond Readvance, though low drumlinoid ridges aligned downvalley are sometimes present, and long, narrow fluted moraines 1–4 m high occupy the floors of some corries, notably Garbh Choire south of Ben More Assynt and the north side of Coire na Creiche in the Cuillin Hills.

**Figure 5.16** Hummocky recessional moraines north of Liathach, Torridon. Each chain of hummocks represents a position of the former glacier terminus as it underwent pulsed retreat. From the valley, these landforms often appear as chaotic hummocks, but viewed from above the pattern of nested ridges is clear.

End moraines, lateral moraines, drift limits and boulder limits have allowed geomorphologists to reconstruct the extent of many Loch Lomond Readvance glaciers (Figs. 4.14 and 4.15), and recessional moraines can be used to track the shrinkage of these glaciers as they retreated towards their sources (Fig. 5.16). Reconstruction of the dimensions of the higher parts of these glaciers is based on different geomorphological evidence. In many locations the upper margin of the ice is recorded by the heads of gullies cut in glacial drift (Fig. 5.17a), and in others it is marked by a *trimline*, where glacial erosion has eroded or 'trimmed' a pre-existing cover of frost-weathered debris (Fig. 5.17b).

Collectively, such evidence has permitted detailed reconstruction of the dimensions of the last Scottish glaciers, enabling geomorphologists to infer the nature of climatic conditions when these glaciers reached their maximum extent. For example, the equilibrium line altitude (ELA) of the icefield that formed during the Loch Lomond Stade on Skye (Fig. 4.15) was about

**Figure 5.17** (a) The upper limit of gullied drift on A'Chioch (947 m) in Kintail marks the former maximum level of a Loch Lomond Readvance glacier. The bulge in the centre of the slope is postglacial rock-slope deformation. (b) A trimline at 700 m altitude (dashed line) cut across Sgùrr Gaorsaic (839 m) in Kintail defines the upper limit of Loch Lomond Readvance glacier ice.

277 m. Mean July sea-level temperatures on Skye, reconstructed from chironomid assemblages, fell to about 6°C at this time. Because the ELAs of glaciers can be linked to particular combinations of precipitation (mainly snowfall) and summer temperature, it appears that at the time the Skye Icefield reached its maximum extent at 12.5–12.4 ka, precipitation on the island was only 60–80% of present-day precipitation. This reduction in precipitation was probably due to the development of extensive sea-ice cover in the North Atlantic Ocean during the Loch Lomond Stade, so that Atlantic airmasses carried less moisture than now.

## Glacifluvial landforms and deposits

The term *glacifluvial* (sometimes rendered as *glaciofluvial* or *fluvioglacial*) is used to describe the landforms created by glacier meltwater streams, and the deposits laid down by glacier-fed rivers. During the summer months, melting snow and ice feeds powerful meltwater runoff, particularly during periods of glacier thinning and retreat. As we saw in Chapter 4, meltwater streams on cold-ice glaciers and ice sheets (those where the ice temperature is below freezing) tend to be restricted to the surface of the ice (*supraglacial* streams) or form *ice-marginal* streams that flow along the glacier margins or *proglacial* rivers that carry drainage away from glacier termini. Warm-ice glaciers where the ice temperature is at or close to 0°C have more complicated plumbing systems: in addition to the routeways identified above, meltwater can descend into the ice via crevasses or vertical shafts and can flow in tunnels within the ice (*englacial* streams), under glacier margins (*submarginal* streams) or along valley floors under the ice (*subglacial* streams). Where meltwater streams have impinged on the underlying land

surface they have cut *meltwater channels*, and around the margins and termini of former glaciers they deposited stratified sand and gravel deposits in a number of distinctive landforms.

### Meltwater channels

Meltwater channels in Scotland fall into two categories: deeply incised channels cut in bedrock, which represent channels occupied by powerful glacial rivers during successive periods of ice-sheet glaciation and deglaciation; and smaller channels, usually about 3–15 m deep, that were eroded in drift deposits (mainly till) as the last ice sheet thinned and retreated. On low ground the former are represented by deep rock-cut gorges, such as those at Corrieshalloch and the Pass of Killiecrankie, both of which were excavated by meltwater streams but are gradually being deepened by postglacial rivers. Mountainous areas, however, support numerous rock-cut meltwater channels that now lack rivers, notably col channels and spur channels. Both typically take the form of straight or meandering V-shaped notches up to about 40 m deep, usually open at both ends (Fig. 5.18). Some are flanked by bouldery scree, many are carpeted by peat, and a few contain ponds or lochans. Intriguingly, some have 'up-and-down' long profiles, with the channel climbing gradually to its highest point then descending on the far side.

Col channels and spur channels formed when supraglacial or englacial streams were lowered or *superimposed* onto the underlying cols and spurs during ice-sheet thinning. As these streams impinged on the ground surface they continued to cut downwards, forming deep channels in the underlying bedrock. Those channels with up-and-down long profiles demonstrate

**Figure 5.18** Col channels cut by glacial meltwater streams. (**a**) Col channel between Corrieyairack Hill (896 m) and Geal Charn (896 m), southern Monadhliath. (**b**) Col channel at the head of Gleann an t-Slugain, Cairngorms.

that the meltwater streams that incised them were originally enclosed within a tunnel in the ice, allowing meltwater to flow up-gradient under hydraulic pressure then down the far side of the col or spur. As the ice surface descended below the levels of col and spur channels, however, they were abandoned as meltwater routeways and left both high and dry, perched across cols, spurs and plateau margins. Most col and spur channels probably reflect repeated episodes of incision as meltwater draining successive ice sheets followed the same routeways.

The deepest col and spur channels have impressive dimensions. That on the floor of the col separating Sgairneach Mhór (991 m) from the Sow of Atholl (803 m) in the west Drumochter Hills is over a kilometre long and 100 m deep; the meltwater gorge west of Beinn Dubhcharach (689 m) in the northwest Monadhliath is up to 130 m deep and joins an even deeper meltwater channel (Conagleann); and the spectacular spur channel of the Dirc Mhór northwest of The Fara (911 m) forms

a deep, rubble-filled gorge over a kilometre long and up to 120 m deep (Fig. 5.19). Several deep col channels cut across the northern Cairngorms, and southwest from Tomintoul the Water of Caiplich has abandoned its original channel to drain northwards through the Ailnack gorge, a deep rocky slot excavated by glacial meltwater erosion. These stunning canyons represent meltwater incision over several glacial cycles, with rockfall contributing to gorge widening during intervening interglacials.

Across low ground and lower hillslopes, thousands of smaller meltwater channels were eroded into drift deposits by marginal, submarginal, subglacial and proglacial meltwater streams as the last ice sheet thinned and retreated. Most of these channels now form dry valleys, some are occupied by tiny 'underfit' streams, and others form part of the present drainage network. On lower slopes and spurs such channels tend to cut obliquely across the slope: ice-marginal channels tend to descend gradually across slopes, whereas

**Figure 5.19** The Dirc Mhór meltwater gorge, which formed through subglacial meltwater incision under successive ice sheets with gorge widening by rockfall during interglacial periods.

sub-marginal drainage channels often cut more steeply across slopes and are frequently interconnected. A particularly impressive array of such meltwater channels occurs along the slopes of Strathallan between Dunblane and Auchterarder, where a lobe of glacier ice retreated westwards during the final stages of the last ice-sheet glaciation. Along the northern slopes of the glen, successive ice-marginal channels up to 2 km long represent thinning of the glacier surface from ~360 m to below 300 m, and along lower slopes on both sides of the valley dozens of interconnected submarginal drainage channels descend obliquely towards the valley floor, where a subglacial river carried meltwater towards the former glacier terminus.

In the Highlands, meltwater coursing along glacier margins during the retreat of Loch Lomond Readvance glaciers was often focused between successive recessional moraines, forming multiple inter-moraine channels aligned obliquely towards valley floors. Such channels must have formed rapidly, possibly even during a single summer runoff season, before retreat of the glacier terminus re-routed meltwater runoff to a new channel farther upvalley. Examples of these transient meltwater routeways can be seen in upper Glen Garry and upper Glen Truim amid the multiple recessional moraines that mark pulsed retreat of the Loch Lomond Readvance glaciers in Drumochter Pass. Even where recessional moraines are absent, numerous small streams draining hillslopes in the Highlands are diverted on lower slopes to follow channels that cut obliquely across the slope, channels that were originally formed by meltwater streams at a time when glacier ice occupied the adjacent valleys.

## Landforms of glacifluvial deposition

Like present-day glacial meltwater streams, the vigorous rivers that drained the last ice sheet and the glaciers of the Loch Lomond Readvance were heavily freighted with sediment. During periods of low flow this mainly comprised silt particles derived from glacial abrasion of bedrock, but during floods much coarser debris – sand, gravel, cobbles and even boulders – was entrained, transported and ultimately deposited, mainly in the *proglacial zone* downvalley from glacier margins. Unlike till deposits, such glacifluvial deposits are commonly stratified, with layers of rounded stones deposited during floods alternating with sandy beds deposited during waning meltwater discharge. Silty beds are rare in most glacifluvial deposits because silt was carried in suspension even

by low flows, and often ended its journey on the floors of lochs or on the seafloor.

During the retreat of the last ice sheet, huge thicknesses of glacifluvial sands and gravel accumulated on valley floors in the Highlands and Southern Uplands in the form of *outwash deposits*, often tens of metres thick, laid down by proglacial rivers as the ice margin receded upvalley. These deposits formed *valley sandar* (singular *sandur*), gently sloping outwash floodplains crossed by braided rivers with multiple interconnected channels. In most glens, however, rivers have subsequently cut down into the original sandar surfaces, leaving them abandoned as *outwash terraces* up to about 40 m above present floodplains (Chapter 9). Such terraces are prominent landforms in many major valleys, such as those occupied by the Rivers Spey, Feshie, Findhorn, North Esk and Tweed. The 2 km wide terrace occupied by the Moss of Achnacree at the mouth of Loch Etive (Fig. 5.20) and the massive terrace at Corran by Loch Linnhe also represent remnants of formerly more extensive sandar. In a few glens, however, valley sandar have escaped extensive dismemberment, as in lower Strathcarron, where an almost intact peat-covered valley sandur surface descends over 8 km from about 40 m above sea level at Balnacra to terminate in a delta at the head of Loch Carron.

Many Scottish sandar and outwash terraces are pitted with enclosed hollows, typically up to about 10 m deep and 200 m wide. These *kettle holes* represent locations where isolated blocks of ice that were stranded as the ice margin receded became buried or surrounded by accumulating outwash deposits. Subsequent melt of such ice blocks resulted in collapse of the overlying or surrounding sands and gravels into the resultant void to form steep-sided depressions, most of which now contain ponds or lochans. Over twenty flooded kettle holes have formed in the outwash deposits of Strathcarron, for example, and several occur on the outwash terraces at Achnacree (Fig. 5.20) and Corran, as well as those that flank the Feshie and Spey Valleys. Particularly large kettle holes are occupied by lochs on the floor of the Spey Valley, such as Loch Insh (~1.5 km²) and Loch Alvie (~0.9 km²), and nearby Loch Morlich occupies a kettle hole ~1.2 km² in area. Much larger kettle-hole lakes occur in the Midland Valley, where both the Lake of Menteith and Loch Leven by Kinross are thought to occupy depressions formed by melting of ice masses stranded amid accumulating glacial or glacifluvial deposits. If this interpretation is valid, Loch Leven (~14 km²) is by far the largest kettle-hole lake in Scotland.

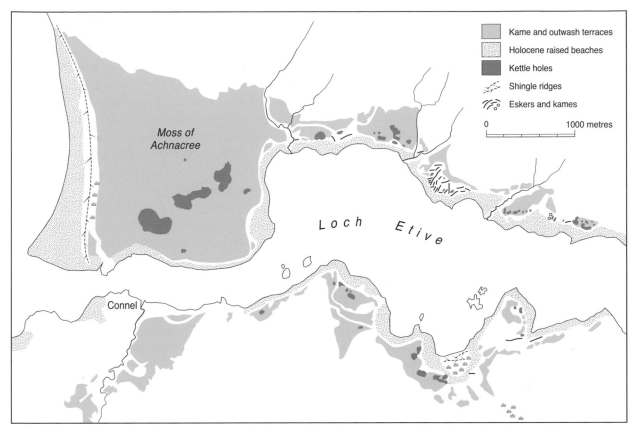

**Figure 5.20** Glacifluvial landforms at the mouth of Loch Etive. Kame terraces rising above the shores of the loch were formed by deposition of sands and gravels along the margins of the glacier that occupied Loch Etive during the Loch Lomond Stade. These merge westwards into the proglacial kettled outwash deposit of the Moss of Achnacree. Adapted from Gray, J.M. (1975) The Loch Lomond Readvance and contemporaneous sea-levels in Loch Etive and neighbouring areas of western Scotland. *Proceedings of the Geologists' Association*, 86, 227–238. © Elsevier 1975, with permission from Elsevier.

In many upland areas, particularly in the Grampian Highlands, rivers flowing along former glacier margins deposited sands and gravels between the edge of the glacier ice and the adjacent valley-side slope. As these glaciers thinned, these deposits were left perched along valley sides as flat-topped *kame terraces* that gradually descend downvalley along the direction of former meltwater flow. Kame terraces occupy hillslopes tens or even hundreds of metres above present valley floors, may be pitted with kettle holes, and are frequently associated with valley-side meltwater channels. Kame terraces that formed at the margins of the last ice sheet are abundant along the flanks of Strathspey and along the northern and eastern slopes of the Cairngorms, where they are perched along slopes and within tributary valleys at altitudes of up to 730 m, and some represent infill of former ice-marginal ponds and lakes by sediment carried by ice-marginal streams. Others are associated with suites of marginal meltwater channels in the upper Clyde,

Annan and Tweed Valleys in the Southern Uplands. In the Streens Gorge of upper Strathdearn, thirteen terrace fragments occur up to 80 m above the present floodplain of the River Findhorn. Here the higher terraces have been interpreted as kame terraces deposited when glacier ice still occupied the gorge, and the lower terraces as outwash terraces. Kame terraces also formed along the margins of some Loch Lomond Readvance glaciers, as in Glen Moriston and along the margins of Loch Etive; at the latter location, kame terraces up to 400 m wide merge westward with the large kettled outwash terrace at Achnacree (Fig. 5.20).

In some areas where outwash deposits (or ablation till deposits) accumulated amid or within stagnating glacier ice, *kame and kettle topography* developed, consisting of flat-topped or conical hillocks (*kames*) separated by numerous kettle holes. Such topography occurs along many upland valleys and on lower slopes, often downslope of arrays of submarginal meltwater channels.

It occupies several locations along the Highland edge northeast of Blairgowrie, notably near the mouth of Glen Esk, is prominent within several of the Angus glens, and occurs along the northern margin of the Southern Uplands. A particularly spectacular area of kame and kettle topography is located on the east side of the River Nith floodplain, 8 km north of Dumfries, and immediately downvalley from multiple subglacial meltwater channels. Kame and kettle topography sometimes mimics that formed by hummocky moraines, and indeed both may be present at locations where glacier ice was cut off from its sources and decayed *in situ*, forming kettle holes amid a chaotic topography of kames composed of outwash sediments and moraines consisting of ablation till.

By contrast, *eskers* – sinuous ridges of sand and gravel originally deposited by meltwater streams in tunnels within or at the base of former glaciers – are rare in most upland parts of Scotland, though outstanding examples occur in some lowland areas such as the Moray Firth coastlands near Inverness. Eskers and kames in lowland Scotland are endangered landforms, as they have proved irresistibly attractive for sand and gravel extraction. The author vividly recalls taking students to examine two landforms previously mapped as eskers. The first had completely disappeared, a victim of gravel extraction. The second turned out to be an abandoned railway embankment.

## Ice-dammed lakes

Glacier ice may act as a dam, impounding an *ice-dammed lake*. This commonly occurs where a glacier occupying a trough dams up a lake in a glacier-free tributary valley, or where a glacier from a tributary valley crosses a glacial trough, damming a lake upvalley. An ice-dammed lake may also form where a glacier advances upvalley against the gradient, impounding a lake in the upper valley. During deglaciation, some lakes burst through or under the ice dam, causing a catastrophic flood known by the Icelandic term *jökulhlaup* ('glacier flood').

The most widespread evidence for former ice-dammed lakes in Scotland consists of laminated silty lake deposits called *varves*. Each varve represents a year's deposition of sediment: during summer, when meltwater is discharging into a lake, relatively coarse silt is deposited, but during winter fine sediment settles from suspension onto the coarse layer. This process is repeated year after year, so by counting the number of varves present it is possible to estimate how long the lake existed. In locations where former ice-dammed lakes rose to a level where the water could drain over a col so that the level of the lake remained constant, horizontal lake shorelines have formed through wave erosion of drift mantling the adjacent slopes. Where streams from tributary valleys drained into an ice-dammed lake they often deposited deltas and fans of sediment that remain perched above valley floors after the lake disappeared.

Varved sediments provide evidence that ice-dammed lakes formed in various lowland areas during retreat of the last ice sheet, but these lakes were transient and left few distinctive landforms. During the Loch Lomond Readvance, however, lakes were dammed by glaciers in several Scottish glens, and their former existence can be traced from the evidence provided by varves, shorelines and perched deltas and fans. One such site is Glen Roy in Lochaber, where we left Agassiz and Buckland formulating the Glacial Theory in October 1840. So well-preserved are the glacilacustrine landforms and sediments of this glen that it has become the most famous geomorphological site in Scotland, subject of over 50 scientific papers and a Mecca for Earth scientists and tourists alike.

## The ice-dammed lakes of Glen Roy and adjacent glens

The most conspicuous manifestation of former ice-dammed lakes in Glen Roy takes the form of three horizontal 'parallel roads' or lake shorelines that can be traced along both sides of the glen at altitudes of 260 m, 325 m and 350 m above sea level (Fig. 5.21). The altitude of each shoreline (and thus the level of the former lake) was determined by that of available outlet cols.

The sequence of shoreline formation in Glen Roy forms part of a wider pattern of ice-dammed lake formation in Lochaber during the Loch Lomond Stade. At this time an outlet glacier fed by the West Highland Icefield advanced into lower Glen Spean, damming a large (>70 km²) lake at 260 m elevation that occupied both Glen Roy and Glen Spean and drained eastwards via a rock gorge near Kinloch Laggan (Fig. 5.22a). At roughly the same time a tongue of ice closed the exit to Glen Gloy, damming a lake that drained over a col at 355 m into upper Glen Roy. Further advance of the Spean Glacier then severed the link between the lakes

**Figure 5.21** The 'parallel roads' (lake shorelines) at the head of Glen Roy. The 260 m shoreline lies at the top of the bluffs in the foreground. The 325 m and 350 m shorelines in the background show how wave erosion steepened the backing slope.

in Glen Roy and Glen Spean, causing the former to rise to 325 m, at which level it drained over a col into the Glen Spean lake near Roughburn. Continued northwards advance of the glacier in Glen Roy then blocked this exit, causing the final rise in lake level to 350 m, at which time the lake drained over a col at this altitude at the head of the glen (Fig. 5.22b). As the ice dam thinned and retreated, the same sequence was repeated in reverse, with the lake level in Glen Roy dropping first from 350 m to 325 m, then from 325 m to 260 m as the lower outlet cols became ice-free. Thus whereas three prominent shorelines are present in Glen Roy, only one (at 260 m) is present in Glen Spean and one (at 355 m) in Glen Gloy.

During the lifetime of the Glen Roy lake, streams draining from tributary valleys deposited large subaqueous fans on the floor of the glen (Chapter 9), and varved lake sediments accumulated on the floors of both Glen Roy and Glen Spean. Detailed research based on the number of varves present in Glen Spean suggests that the Lochaber ice-dammed lakes existed for just over 500 years, from ~12.2 ka to ~11.7 ka. The demise of these lakes was spectacular: bursting of the ice dam in lower Glen Spean released up to 5 km³ of

water in a catastrophic jökulhlaup, an enormous flood that escaped via the Spean gorge to flow northeast along the Great Glen to Inverness, where it deposited a huge offshore gravel fan 7 km² in area. The Loch Ness 'plesiosaur', if it had somehow survived burial under a kilometre or two of glacier ice, would surely have been swept into the Moray Firth as a wall of water swept up the Great Glen.

### Other ice-dammed lakes

Varved deposits, shorelines and deltas provide evidence that the glaciers of the Loch Lomond Readvance dammed lakes in several other glens. Near Bridge of Orchy, five faint shorelines in 248–332 m altitude record the progressive lowering of a large ice-dammed lake up to 155 m deep that formerly occupied the valley east of Loch Tulla, each shoreline being related to that of outlet cols that became ice-free as the glacier margin retreated westwards. At both the northern and southern shores of Loch Garry near Drumochter Pass are raised deltas that formed when glaciers dammed the loch at both ends, raising it about 20 m above its present level. In Wester Ross, a tongue of ice from the Fannich Mountains crossed Strath Bran, damming a

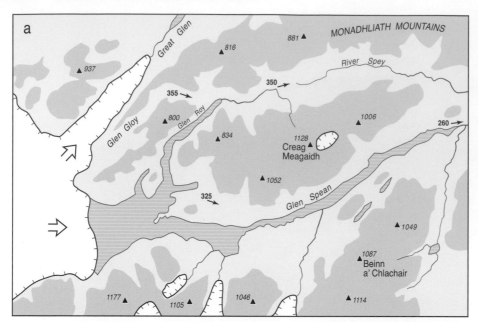

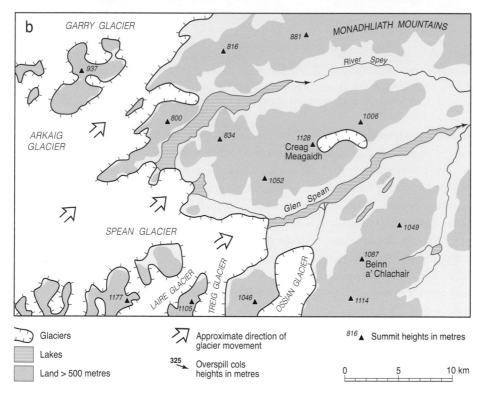

**Figure 5.22** Ice-dammed lake formation in Glen Roy and Glen Spean. (**a**) Glacier ice advancing into Glen Spean from the west dams a lake that drains westwards over a col at 260 m. (**b**) Further advance of the Spean Glacier severs the Glen Roy lake from the Spean Lake, forcing the Glen Roy lake to rise first to 325 m (draining across a col to the southeast) then to 350 m (draining westward into the Spey).

lake that flooded the Achnasheen area to a depth of about 130 m. Here the most impressive evidence consists of two spectacular deltas deposited at Achnasheen by glaciers that advanced into the lake from the west and southwest, now represented by broad, gently sloping terraces about 20 m above the present floodplain near the village. Finally, along the northern slopes of Glen Doe, seven shorelines record the existence of an ice-dammed lake up to 135 m deep that was dammed by a tongue of ice that extended eastwards down neighbouring Glen Moriston; belts of water-scoured bedrock downvalley from this former lake suggest that it drained catastrophically along the margins of and underneath the damming glacier. Other ancient lakes dammed in tributary valleys by Loch Lomond Readvance glaciers await discovery and documentation.

## Synthesis: glacial landscapes of Scotland

As will be obvious from the foregoing account, there are marked differences in the glacial landform assemblages present in different parts of Scotland. Various attempts have been made to zone the Scottish landscape in terms of the degree of landscape modification by glacier ice or the dominant glacial landforms present. This (rather subjective) exercise depends on the criteria used to identify particular terrain types and their boundaries, but the scheme outlined in Figure 5.23 is fairly similar to earlier versions. Inevitably, painting on a large canvas with a broad brush tends to obscure details, but nevertheless highlights some of the major topographic contrasts that lend Scottish landscapes their astonishing geodiversity.

The most dramatic mountain scenery in Scotland falls within the zone of *alpine glacial landscapes*. These occupy a large part of the western Highlands, with outliers on North Harris, the Cuillin Hills of Skye, the mountains of Rum, central Mull and north Arran, and the Galloway Hills. These are areas of advanced glacial modification of resistant igneous or metamorphic rocks, characterized by deep glacial troughs, glacial breaches and rock basins, with mountain ridges indented by corries and very limited preservation of palaeosurfaces. The distribution of alpine landscapes is similar to the extent of glacier ice during the Loch Lomond Readvance, suggesting that valleys within this zone were re-occupied on numerous occasions throughout the Pleistocene by warm-ice-based

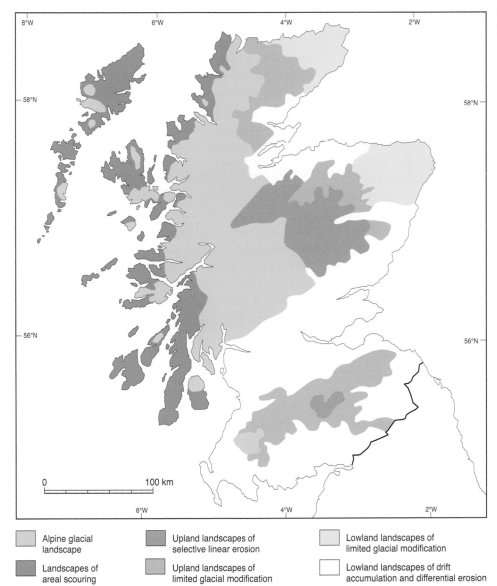

**Figure 5.23** Glaciated landscapes of Scotland.

Alpine glacial landscape

Landscapes of areal scouring

Upland landscapes of selective linear erosion

Upland landscapes of limited glacial modification

Lowland landscapes of limited glacial modification

Lowland landscapes of drift accumulation and differential erosion corrie and valley glaciers as well as by fast-flowing ice within successive ice sheets. Because of its close correspondence with the extent of Loch Lomond Readvance glaciers (Fig. 4.14), this zone also hosts numerous end, lateral, recessional and hummocky moraines.

Flanking the alpine zone to the west are *landscapes of areal scouring*, formed mainly by the erosion of resistant rocks during episodes of ice-sheet glaciation. The hallmarks of this zone are wide expanses of glacially moulded bedrock, often partly buried by peat, numerous whalebacks and roches moutonnées, and areas of knock-and-lochan terrain, particularly on the gneiss of the Outer Hebrides and the peninsulas of the western seaboard north of Applecross. Aerial scouring is also well represented on the lava terrains of Skye and Mull and on the schists that underlie the peninsulas between Loch Duich and the Sound of Mull, locally extending to the summits of mountains in these areas.

Much of the Eastern Grampian Highlands forms a zone of *selective linear erosion*. This type of landscape is epitomized by the Cairngorms, the Monadhliath Mountains, the Gaick plateau, much of the Southeast Grampians and the Tweedsmuir Hills in the Southern Uplands. In these areas palaeosurfaces have been preserved as gently rolling plateaux but are incised by deep Icelandic troughs such as Glen Avon and Glen Clova. Corries are numerous along plateau margins and at trough-head locations. During the Loch Lomond Readvance some of these areas supported plateau ice caps and corrie glaciers, but these rarely reached low elevations. This zone also supports numerous meltwater channels and some of the finest outwash terraces in Scotland, but moraines are mainly confined to corries and high-level valleys.

The contrast between the alpine landscapes of the western Highlands and the landscapes of linear erosion farther east is thought to reflect several factors. It is likely that the western Highlands were strongly dissected by rivers prior to the onset of glaciation, as a result of earlier uplift and tilting. Moreover, the mountains of the western Highlands probably experienced repeated episodes of icefield and valley glaciation (similar to the Loch Lomond Readvance) at times when the eastern Grampians supported only limited glacier growth. Finally (and probably most importantly), the contrast between the Alpine landscapes of the western Highlands and the plateau landscapes of the eastern Grampians

and Cairngorms is attributable to eastwards reduction in snowfall during successive periods of glacier expansion and ice-sheet growth. In the west, large inputs of snowfall from Atlantic airmasses resulted in rapid accumulation of glacier ice, so glaciers in this zone tended to be fast-flowing, warm-ice-based and erosive. On the snow-starved mountains of the east, ice accumulation was slower and plateau ice caps were predominantly cold-ice-based, frozen to the underlying substrate and hence capable of only very limited erosion. Only in deep valleys and corries did the base of the ice reach melting point, enabling it to widen and deepen glacial troughs and excavate corrie floors – hence the highly selective nature of glacial erosion in such terrain.

Much of the Southern Uplands and the Grampians northeast of the Cairngorms forms *uplands of limited glacial modification*. These are zones of relatively low (< 700 m), rolling hills and plateaux where palaeosurface fragments are often preserved and landforms of glacial erosion such as troughs, rock basins and corries are rare or poorly developed, probably because such areas were occupied by cold-based, non-erosive ice for much of the lifetime of the ice sheets that crossed them, though meltwater channels are present on many hillslopes and outwash terraces occupy low ground. The Buchan area of northeast Scotland also represents a *lowland landscape of limited glacial modification*, where preglacial basins, palaeosurface fragments and saprolite covers are preserved. Much of Caithness falls into a similar category.

Finally, the Midland Valley represents a classic area of *drift accumulation and differential glacial erosion*, being blanketed for the most part in thick accumulations of glacial and glacifluvial deposits, above which resistant igneous outcrops (lavas, volcanic plugs and sills) form the high ground. The Solway lowlands form similar terrain, as does the lower Tweed Valley, though the latter lacks igneous protuberances.

As the last ice sheet thinned and retreated, and subsequently during the Loch Lomond Stade, much of Scotland remained outside the reach of glacier ice, and mountain summits protruded through the ice as nunataks. At such times, glacier-free areas were exposed to extremely cold periglacial conditions dominated by deep freezing of the ground. Even today, the higher parts of Scottish Mountains fall within the periglacial realm. The distinctive mountain landforms produced under such conditions form the topic of the next chapter.

# Chapter 6

# Periglacial landforms

## Introduction

For many hillwalkers, the best part of a day in the Scottish mountains is when they emerge onto high ridges or plateaux, with the next summit in view and spectacular vistas all around. But visitors to the high summits cannot fail to notice that there is often a marked change from the slopes they have just ascended to distinctive landscapes without parallel on lower ground. This is because they have entered the *periglacial zone*, where the dominant geomorphological process has been *frost action*, the recurrent freezing and thawing of the ground.

The term *periglacial* is used to describe the climatic conditions, processes, landforms, sediments and soil structures associated with cold nonglacial environments on Earth (and on Mars, where periglacial landforms have also been detected). Periglacial environments presently occupy about 25% of the Earth's land area and coincide approximately with areas where the mean annual air temperature is below +3°C. They presently encompass not only vast swathes of the Arctic and Subarctic as well as ice-free parts of the Antarctic and Subantarctic islands, but also the higher parts of mid-latitude mountains. Moreover, during glacial stages as the great Quaternary ice sheets expanded, periglacial conditions extended far beyond the ice-sheet margins. Many European landscapes from southern England through France, Germany and Poland to Russia were largely fashioned by periglacial processes operating during successive glacial stages.

In areas where mean annual air temperatures are below -1°C, some or all of the ground is underlain by *permafrost*, perennially frozen ground that does not thaw during the summer months. Permafrost is overlain by the *active layer*, a zone typically 0.3–2.0 m thick that thaws each summer then re-freezes each winter. The boundary between the active layer and the underlying permafrost is called the *permafrost table*. Just below the permafrost table the upper parts of permafrost often contain high concentrations of *ground ice*, usually in the form of *pore ice*, which occupies voids in

soil or rock, and *ice lenses*, which are thin platelets of ice formed during slow freezing of the ground. Even in the absence of permafrost, winter freezing of fine-grained soils results in the formation of ice lenses as soil water is drawn upwards to the descending freezing plane. The formation of ice lenses causes an increase in soil volume, resulting in *frost heave*, a powerful process that you may be familiar with through disruption of tarmac roads, upheaving of boulders in ploughed fields and even buckling of poorly constructed patios.

Permafrost presently underlies ~16% of the global land area, but during glacial stages it extended far beyond its present limits. We know this because certain landforms and soil structures develop only in permafrost, and their remnants are detectable in areas that now experience temperate climates. The most common features diagnostic of former permafrost are *ice-wedge pseudomorphs* (sometimes called *ice-wedge casts*), which are commonly found in gravel pits, including some in Scotland (Fig. 6.1). Their origin is intriguing. In areas underlain by permafrost, a sharp reduction in winter air temperature causes the frozen ground to contract and crack open, forming a network of *frost polygons* commonly 5–50 m wide. In spring, snow meltwater enters the cracks and freezes, eventually forming a V-shaped downward-tapering *ice wedge* under polygon margins. When the climate warms at the end of a glacial stage and ice wedges melt, sediment fills the void, forming distinctive V-shaped structures – ice-wedge pseudomorphs – that interrupt the bedding in sand and gravel deposits. The former frost polygons can also sometimes be detected on aerial photographs and even on satellite imagery. Thus permafrost, like glacier ice, leaves unmistakable traces in the landscape that can be detected millennia after it has disappeared.

On Scotland's mountains, periglacial features fall into three groups: (1) relict periglacial landforms that pre-date the growth and demise of the last ice sheet;

**Figure 6.5** Relict Lateglacial periglacial landforms. (**a**) Large-scale relict sorted circles (vegetated areas are cells of fine sediment 2–3 m wide) on a blockfield near the summit of Glas Maol, southeast Grampians. (**b**) Earth hummocks merging into relief stripes on nearby Cairn of Claise. (**c**) Meandering relief stripes on Seana Braigh, northwest Highlands. (**d**) Stone-banked solifluction lobes on Creag an Leth-choin in the Cairngorms.

blockfields and upper debris-mantled slopes. On level or gently sloping ground these now take the form of vegetation-covered cells of soil surrounded by boulders (Fig. 6.5a), and on slopes they consist of downslope-aligned bands of boulders that alternate with gravelly or vegetation-covered soil. Relict nonsorted patterned ground on vegetated plateaux commonly takes the form of *earth hummocks*, vegetated mounds of soil up to about 0.5 m high and 2.0 m in diameter separated by a network of depressions (Fig. 6.5b). As the slope steepens, these sometimes grade into *relief stripes*, alternating vegetated ridges and furrows aligned downslope (Fig. 6.5c). Earth hummocks are common on vegetation-covered plateaus, particularly on schists, but relief stripes are rarer; the finest examples are those on the slopes below Tom a'Chòinnich (953 m), a subsidiary summit of Ben Wyvis.

The formation of both sorted and nonsorted patterned ground is due to *differential frost heave* of the ground.

Ground subject to winter frost heaving through the formation of ice lenses in the soil initially experiences fairly uniform uplift. Over multiple annual freeze–thaw cycles, however, a preferred 'wavelength' of frost heave develops in freezing and expanding soil, so that regularly spaced domes develop. Where clasts are abundant, these are frost-heaved to the surface and migrate to the margins of such domes, forming sorted circles. Where clasts are less abundant, the domes evolve into hummocks through upward movement of soil in the developing hummock and compensatory inward migration of soil from hummock margins. On slopes, the movement of clasts and soil is also affected by gravity-induced downslope movement, so that regularly spaced sorted stripes or relief stripes form through time.

The second effect of annual freezing and thawing of the active layer during Lateglacial cold periods was *solifluction*, defined as the slow downslope movement of the soil due to repeated freezing and thawing of the ground. On slopes underlain by permafrost, solifluction comprises three processes: *frost creep*, *gelifluction* and *plug-like deformation*. Frost creep is caused by recurrent annual frost heave and thaw re-settling of the ground: because heave pushes the ground up normal (at 90°) to the slope, but the ground settles near-vertically on thawing, there is net downslope movement of the ground with each annual freeze–thaw (heaving–settling) cycle. Gelifluction occurs during thaw of the active layer. As the soil thaws, melt of ice lenses releases water, but because the rate at which water is released exceeds the rate at which it can escape towards the ground surface, high *pore-water pressures* are generated. These 'soften' the soil just above the thaw plane, causing it to deform downslope. As thaw is from the surface downwards, this effect begins near the surface and gradually extends downwards as thaw descends into the soil, but dies out at depths of about 0.5–0.6 m. Plug-like deformation describes the *en masse* movement of soil as ice lenses at the base of the active layer melt, generating high pore-water pressures that cause the soil just above the permafrost table to deform, carrying the overlying active layer gradually downslope. Over a single annual freeze–thaw cycle, these three processes collectively result in downslope ground surface movement of about 10–50 mm a$^{-1}$, with velocity diminishing with depth. This may not seem much, but over a millennium it implies downslope displacement of the ground surface of 10–50 m.

The most impressive Lateglacial landforms produced by solifluction are large terraces (on gentle upper slopes) and lobes (on steeper slopes) composed of boulders. These landforms are best developed on debris-mantled slopes underlain by granite, particularly in the Cairngorms and on Lochnagar, and are known as *stone-banked terraces* and *stone-banked lobes* (Fig. 6.5d). The relatively steep fronts or *risers* of these landforms are usually up to 3 m high, and the steps or *treads* upslope of the risers are usually covered by boulders and vegetation. Individual lobes are typically 10–25 m wide. It is thought that these terraces and lobes represent downslope movement of frost-heaved boulders that have been gradually rafted downslope by solifluction operating in the underlying soil. Smaller stone-banked terraces and lobes occur on upper slopes underlain by other rock types, particularly quartzite and some schists. In the Cairngorms, stone-banked lobes descend to 540 m, but are absent from ground that was occupied by glaciers during the Loch Lomond Stade, suggesting that they ceased to move

as temperatures rose rapidly at the beginning of the Holocene, causing the underlying permafrost to thaw.

Other relict Lateglacial solifluction lobes are completely covered by vegetation. These commonly occur on lower slopes where the gradient lessens and are often about two metres thick, but with degraded, gently sloping risers. Even where lobes are absent, Lateglacial solifluction has played a major role in modifying the form of lower slopes in many mountain areas. Outside of the limits of the Loch Lomond Readvance glaciers, such slopes are commonly mantled by *soliflucted till sheets*: the till was deposited by the last ice sheet, but moved slowly downslope by Lateglacial solifluction, forming smooth, featureless hillslopes. The lower slopes of mountains in the Southern Uplands are extensively mantled by soliflucted till sheets, as are lower slopes throughout much of the central and eastern Grampians.

Lateglacial solifluction has also played a role in the accumulation of thick *periglacial valley-fill deposits*, particularly in the Southern Uplands. These take the form of low-gradient (3–15°) valley-floor drift benches, 20–300 m wide, with steep frontal bluffs 3–20 m high formed by later river incision. These benches are typically composed of *in situ* till deposits overlain by till and frost-weathered debris that has been transported down adjacent valley slopes and deposited on valley floors. Whether solifluction has been the sole agent of downslope sediment transport is debatable, as the thickness of some valley-fill deposits suggests that a more rapid process may have been involved. One candidate is *active-layer detachment sliding*, whereby rapid thaw of ice lenses at the base of the active layer generates exceptionally high pore-water pressures that destabilize the active layer, causing it to slide or flow downslope, even over low gradients. Active layer detachment slides are increasingly common in present permafrost areas as the climate warms, but their possible contribution to downslope transport of sediment in Scotland during (and probably at the end of) Lateglacial cold periods remains to be established.

## Active periglacial landforms

The higher parts of Scottish mountains presently experience a *maritime periglacial climate*, characterized by strong winds and high precipitation rather than severe cold or deep ground freezing. Mean annual air temperatures recorded over the period 1981–2010 for the summits of Cairngorm (1245 m), Aonach Mór (1130 m) and The Cairnwell (933 m) were +0.9°C, +1.7°C and

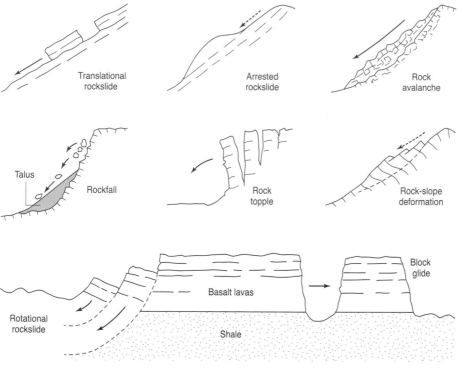

**Figure 7.3** Types of rock-slope failure. Rotational rockslides and block glides are mainly confined to the lavas of the Hebridean Igneous Province.

**Figure 7.4** Catastrophic rock-slope failures in the Scottish Highlands. (**a**) An arrested rockslide near the summit of Meall na Dige in the Crianlarich Hills. Winter sunlight highlights two large blocks of displaced rock. (**b**) The Beinn Shiantaidh rock avalanche, Jura, where rapidly moving landslide debris rebounded at the base of the slope to form a large arcuate ridge.

but large rockfalls are rare. *Toppling failures* occur when a steep slab of rock tilts away from a cliff face and topples forwards down the slope; many large-scale rockfalls from steep cliffs are toppling failures. The products of rockfall accumulate progressively at the foot of cliffs to form aprons of debris referred to as *talus slopes* or *scree slopes*.

The final category of rock-slope failure takes the form of deep-seated downslope movements of entire mountainsides, but without detachment of blocks or runout of debris: the mountainside shifts downslope, but

movement is then arrested and stability regained. These landslides are usually termed *rock-slope deformations* or *deep-seated gravitational slope deformations*. Many of the largest (>0.25 km$^2$) rock-slope failures in Scotland are of this type.

Many landslides, however, exhibit characteristics of more than one type. Catastrophic failures often involve a combination of sliding and toppling, though the former usually predominates, and some rock-slope deformations exhibit localized collapse. Some types are transitional:

there is, for example, no clear morphological distinction between rock-slope deformations and large arrested slides with limited downslope displacement of rock. Moreover, as most large rock-slope failures in Scotland occurred more than 10,000 years ago, it is sometimes difficult to reconstruct the mode of failure from the surviving geomorphological evidence.

## Causes of rock-slope failure

Geomorphologists identify three inter-related sets of factors that ultimately lead to rock-slope failure: preconditioning factors, which represent the intrinsic state of a rock slope; preparatory factors, which cause weakening of the slope; and trigger factors, those responsible for initiating movement. The most important preconditioning factors are those related to the type and structure of the underlying rock. Some intact rocks, such as granite, are much stronger than others, such as shales. However, the most important preconditioning rock properties are often rock structure and jointing. Rocks with bedding planes dipping steeply towards the slope, for example, are more likely to experience sliding failure

than those where bedding planes are horizontal or dip into the slope. The most critical features, however, are the depth, density and alignment of *joints* (fractures) in a rock mass (Fig. 7.5). All near-surface rocks contain joints, though in some cases these are widely spaced and in others they are densely distributed throughout the rock. Joints are formed in various ways: by tectonic stress, unloading, and (in some igneous rocks) by contraction of rocks as they cooled. Joints form zones of weakness within rock masses, particularly when they are interconnected to form a potential failure plane. In such circumstances the stability of a rock slope is determined not by the intact strength of the rock itself, but by internal rock bridges, interlocking protrusions across joints, and the frictional strength of the contacts between the rocks on either side of joints.

By comparing Figure 7.2 with a geological map (Fig. 2.1) it can be seen that the great majority of rock-slope failures on the mainland north of the Highland boundary have occurred in metasedimentary rocks, particularly schists; they are comparatively rare on slopes underlain by gneiss, sandstone or granite. South of the Highland

**Figure 7.5** (**a**) Stress-release joints in granite. Dilation joints parallel to the cliff face are crossed by low-angle joints that become more closely spaced towards the crest of cliffs. (**b**) Vertical fissures formed by opening of joints at the headwall of a rock-slope failure on Beinn Narnain, Argyll.

## Rockfall

Rockfall is caused by several processes. These include progressive weakening of cliffs by paraglacial stress release through extension and opening of joints, together with earthquakes and build-up of water pressure in rock joints following prolonged rainstorms. Chemical weathering may weaken frictional contacts along joints, and even heating and cooling of the rock face, which causes very slight expansion and contraction of exposed rock, may loosen and detach unstable blocks. A particularly important cause of rockfall in cold environments is freeze–thaw activity: water in joints near the cliff face freezes, and as the volume of ice is 9% greater than that of water, the joint is prised open. Over multiple freeze–thaw cycles this process may gradually destabilize blocks of rock on a cliff face, until a final freezing episode delivers the *coup de grâce*, and the ensuing thaw releases the block onto the talus below.

Rockfall activity is measured as *rockwall retreat rate*, which averages the effects of rockfall across an entire cliff face. Long-term rates of rockwall retreat (in metres per millennium) are measured by calculating the volume of talus at the foot of a cliff and dividing by the area of the contributing cliff and the period of talus accumulation. In Scotland, measurements based on the volume of talus below the basalt cliffs of the Trotternish escarpment on Skye suggest that during the Lateglacial period following ice-sheet deglaciation (~16.5–11.7 ka) the cliff retreated on average about 3.9 m due to intermittent rockfalls, implying a rockwall retreat rate of 0.6–1.0 m ka⁻¹. Other studies suggest that under the severe periglacial conditions of the Loch Lomond Stade cliff retreat at various sites in the Highlands ranged from ~1.6 m ka⁻¹ to ~3.8 m ka⁻¹, reflecting the combined effects of paraglacial stress relief and deep annual freeze–thaw cycles at that time. Formation of talus accumulations below cliffs inside the limits of Loch Lomond Readvance glaciers demonstrates that at such sites rockfall remained active during the early Holocene prior to about 7.0 ka, but radiocarbon dating of buried organic soils within talus accumulations suggests that only sporadic rockfall has occurred since then.

However, rockfall has not been the only process contributing to cliff retreat after deglaciation. Talus accumulations comprise not just rockfall boulders, but also an infill of silt, sand and fine gravel that represents weathering and detachment of small particles that have fallen from the source cliffs and been washed by rain into the talus accumulations downslope. This fine material makes up to 27–30% by volume of talus, implying that weathering and detachment of small particles may have contributed to almost a third of overall cliff retreat since deglaciation.

## Talus accumulations

The accumulation of rockfall debris at the foot of cliffs has resulted in the formation of *talus slopes* that consist of a steep (34–38°) straight slope with a short basal concavity. Talus slopes take three forms: *talus sheets*, where rockfall input along the foot of cliffs has been fairly uniform; *talus cones*, which result from concentration of rockfall debris below large rock gullies and *coalescing talus cones*, formed by the lateral merging of individual cones (Fig. 7.13). It is thought that the concavity at the foot of talus slopes develops first, when the full length of the rockwall is exposed so that falling debris has high kinetic energy, allowing large boulders to travel far from the cliff foot. As the talus accumulates against the cliff, however, the average height of fall of boulders is reduced, so they tend to come to rest farther up the slope, usually amongst debris of similar calibre. As a result, the slope gradually steepens and extends over the basal concavity, and develops *fall sorting*, with the largest boulders accumulating on the footslope and the smaller ones becoming trapped on the upper slope (Fig. 7.13a).

Most Scottish talus slopes, however, are essentially relict landforms that accumulated during the Lateglacial period. The majority are partly or completely vegetated, such as those that skirt the Trotternish escarpment and the eastern slopes of Ben Meabost on Skye. Unvegetated talus slopes exist only where the debris has been too coarse to permit the establishment of soil or vegetation cover, such as those in Coire Làgan in the Cuillin Hills, or at sites where recent rockfalls have spread debris over the vegetation cover (Fig. 7.13). Moreover, many talus slopes are now deeply scarred by gullies, which indicate that erosion has replaced accumulation as the dominant form of geomorphic activity. Radiocarbon dating of buried soil horizons exposed in gullies suggests that most Scottish talus slopes reached their present dimensions within five millennia after deglaciation, thereafter becoming progressively eroded by gullying. Various processes have contributed to gully formation in talus, notably shallow sliding of talus debris, debris flows (see below) and stream erosion during exceptional rainstorms. Sediment excavated from gullies by debris flows has often accumulated at the foot of talus slopes in the form of debris cones. Figure 7.14 illustrates the life cycle

**Figure 7.13** (**a**) Talus cones formed by accumulation of rockfall debris in Coire Làgan, Cuillin Hills, Skye, showing the downslope increase in boulder size. The scar running down the cone on the left is the route taken by descending climbers. (**b**) Vegetation-covered relict talus slopes below cliffs in Coire a'Mhuillin, An Teallach. The ridge crossing the valley floor is the end moraine deposited by the glacier that occupied Coire a'Mhuillin during the Loch Lomond Stade.

of typical talus slopes in Scotland, from accumulation in the wake of retreating glacier ice to maturity and eventual dismemberment by gully erosion.

## Snow avalanche landforms

Snow avalanches in arctic and alpine mountains play an important role in eroding bedrock gullies (*avalanche* *chutes*) near the crests of cliffs and in modifying talus slopes by stripping debris from upper slopes and redepositing it farther downslope in the form of a long sweeping concavity. The most distinctive landforms produced in this way are *avalanche boulder tongues*, embankments of reworked talus debris that extend across valleys. Both relict (Lateglacial) and active avalanche

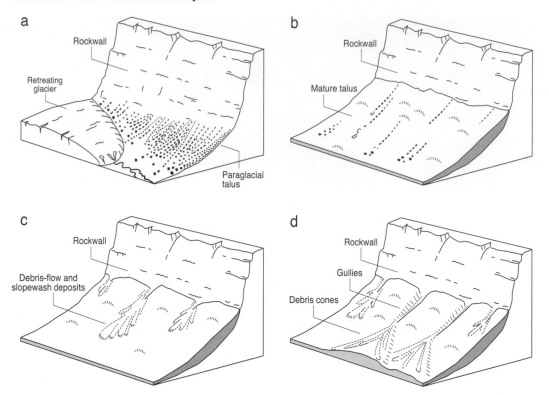

**Figure 7.14** Typical evolution of talus slopes in Scotland. (**a**) Stress release and frost action cause frequent rockfalls and rapid talus accumulation after glacier retreat. (**b**) After a few millennia, the rockwall has stabilized; rockfall has diminished; soil and vegetation cover form on the talus. (**c**) Gully incision at the talus crest, and redeposition of sediment by debris flows on the upper slope. (**d**) Debris flows and runoff within gullies cause gully extension and deposition of sediments as slope-foot debris cones. From Hinchliffe, S. and Ballantyne, C.K. (2009) Talus structure and evolution on sandstone mountains in NW Scotland. *The Holocene*, 19 (3), pp. 477–486. © 2009 SAGE publications.

landforms are present on Scottish mountains, though neither are widespread.

Typical examples of relict avalanche boulder tongues occur downslope from avalanche chutes at 910–970 m near the head of Glen Luibeg in the Cairngorms. Also in the Cairngorms, upper-slope rock gullies at the crest of a northeast-facing slope below Creag Mhigeachaidh closely resemble active avalanche chutes in Spitsbergen, and similar landforms may be more widespread but have escaped identification. Relict avalanche boulder tongues occur at a few locations elsewhere in the Highlands, for example on the floor of upper Glen Feshie below Creag na Gaibhre and below the cliffs of Beinn Dearg Mheadhonach on Skye. The apparent paucity of evidence for Lateglacial snow avalanche activity in Scotland may be accounted for by destruction or burial of such evidence by Holocene slope processes, notably debris flows.

Under present conditions, wet-snow avalanches, shallow slab avalanches and cornice collapses are fairly common in Scotland, particularly after blizzards or during thaw periods. Most begin on steep (30–45°) upper slopes where windblown snow has built up in the lee of plateaux or ridges. Because periods of thaw often interrupt snow accumulation, most avalanches are small and few have a significant geomorphic impact. This is mainly because the majority travel across snowcover, and therefore neither entrain nor deposit debris, though occasional 'dirty' snow avalanches can be observed, usually in spring. In many areas, evidence for recent snow avalanche activity is confined to uprooting of turf, localized erosion of soil and deposition of a spread of soil and stones downslope. After snowmelt, these minor effects can be identified by 'perched' clasts and soil resting on boulders or vegetation.

Active avalanche boulder tongues are rare, though several occur in the Lairig Ghru, the through valley that bisects the Cairngorms. The most spectacular are those that dam the Pools of Dee near the highest part of the pass, about 800 m above sea level. Here successive debris-freighted snow avalanches have swept down the steep eastern slopes, depositing long tongues of debris up to 7 m thick that cross the valley and extend a short distance up the opposite slope. The freshness of some of the clasts on these boulder tongues indicates that though these impressive landforms may have accumulated over millennia, they have been refreshed by recent additions of snow avalanche debris.

Another effect of snow avalanche activity is the formation of *avalanche impact pits* that develop at sites where successive avalanches impact a basal break of slope on a corrie or valley floor, excavating a pit and ejecting debris that forms an *avalanche impact rampart* on the far side of the pit. A superb example is perched at 910 m below the cliffs on the northwest side of Ben Nevis in Coire na Ciste, where the depression now occupied by Lochan Coire na Ciste has been excavated by the impact of successive powerful avalanches from the head of the corrie. The ejected debris forms a 75 m long avalanche impact rampart that dams the lochan, with a spread of angular boulders on its far side. Much of the debris is fresh, and numerous clasts are 'perched' on larger boulders, a hallmark of avalanche deposition that results from clasts being lowered onto underlying boulders as the snow melts.

## Slope failures in soils and peat

A common form of slope failure on Scottish mountainsides takes the form of shallow translational slides, where a slab of soil, typically 0.4–2.0 m thick, ruptures from the surrounding soil and slides downslope, often breaking up into smaller soil blocks or a muddy slurry as it does so. These superficial landslides generally occur on slopes of 25–35° and are common on slopes mantled by till, particularly on the sides of steep bluffs where streams have incised till deposits (Fig. 7.15). Recent slides leave disfiguring scars, each comprising a steep backscarp, sidescarps of soil and roots, a slope-parallel failure plane, detached blocks of soil and vegetation, and sometimes a lobe of rumpled soil that has been pushed into ridges by movement of soil from upslope. Some translational slides occur within the soil mantle, others over the underlying bedrock. Over a decade or two, however, the scars degrade and vegetation colonizes failure sites, leaving only indistinct hollows on hillslopes. Sometimes bedrock remains exposed in the scar, or springs fed by groundwater emerge at the surface, allowing a flush of brilliant green moss to colonize the hollow.

**Figure 7.15** Cuspate scars representing former translational landslides in drift deposits, Glen Artney, southern Grampians. The scars are about 40 m wide.

Shallow translational landslides are triggered by exceptionally prolonged, intense rainstorms, particularly if the ground is already saturated by previous rainfall or snowmelt. Rupture occurs when the shearing force operating on the slope exceeds the shearing resistance of the soil. The shearing force is simply the downslope component of the weight of the soil ($=W \sin \alpha$, where W is the weight of the soil and $\alpha$ is the slope gradient). The shearing resistance is determined by several factors, but the most important in determining slope stability is the 'normal' component of soil weight ($=W \cos \alpha$), which binds the soil to the slope. As the water table rises in the soil during rainstorms, the saturated soil below the water table becomes partly buoyant, and rising pore-water pressures (u) reduce the effective 'normal' component of soil weight ($=W \cos \alpha - u$) that holds the soil in place. If the pore-water pressure rises sufficiently, the shearing resistance of the soil falls below the shearing force and a slab of soil ruptures and moves downslope. Sometimes stability is regained as water drains from the mobile soil, leaving little more than tension cracks or a slightly displaced slab of soil; sometimes the slab travels downslope more or less intact; and in some cases the sliding slab starts to disintegrate and flow, as described in the next section.

A related phenomenon is the occurrence of *peat slides*, which involve sliding of intact bodies of peat over bedrock or the underlying mineral soil, sometimes over slopes as gentle as 6°. Peat slides are also triggered by extreme rainstorms and consequent build-up of pore-water pressure at the base of the peat; this reduces the shearing resistance, causing a raft of peat to move downslope. Documented examples in Scotland have involved peat up to 3.0 m thick and peat bodies up to 72,500 m² in area, though 5000–10,000 m² is more typical. The resultant landforms include arcuate or irregular failure scars, intact rafts of displaced peat, fragmented peat blocks and sometimes pressure ridges at the downslope end of the slipped mass. Most reported examples occur on the peat-mantled hills of Shetland, but this probably reflects under-reporting of peat slides in other upland areas.

## Debris flows

The term *debris flow* describes the rapid downslope flow of coarse debris mixed with mud and water and is also used to describe the resultant landforms (Fig. 7.16). Debris flows typically advance downslope at a few metres per second, often in surges, and involve *en masse* flow of saturated debris; water content is usually no more than 10–30% by weight. Two types occur on Scottish mountains: *hillslope debris flows*, which occur on open slopes, and *channelized debris flows*, which follow gullies. The two types are often transitional: flows initiated in gullies can encroach on open slopes and flows originating on open slopes may become focused within gullies. Hillslope flows usually originate on steep (25–40°) slopes mantled by till, talus or weathered regolith; they occur on all rock types but are particularly common on sandy soils derived from granite or sandstone. Channelized debris flows may be initiated on slightly lower gradients. At many locations, deposition of sediment by recurrent debris flows following the same track has resulted in the accumulation of slope-foot *debris cones* (Fig. 7.14). Numerous examples occur along the lower slopes of glacial troughs such as Glen Etive, Glen Coe, Glen Docherty and Gleann Einich. Such debris cones typically have gradients of ~10–25°, less steep than talus cones (34–38°) but steeper than alluvial fans (<10°).

Debris flows occur on steep soil-mantled slopes throughout Scotland. Most individual flows are small, and involve less than 50 m³ of sediment, though a few recent flows have transported more than 1000 m³. The majority follow long, narrow tracks that emanate from upper-slope gullies and are marked by distinctive parallel *levées* (marginal ridges) that terminate in one or more lobes of bouldery debris, though larger flows sometimes spread out over lower slopes, and channelized flows may lack marginal levées.

Debris flows are triggered by prolonged rainstorms of exceptional intensity. Torrential rainfall over Lochaber in 1953 caused multiple debris flows that damaged roads, culverts and forestry. A rainstorm of >80 mm in 24 hours triggered 71 debris flows in the Lairig Ghru in 1978, and another in 2004 generated 31 debris flows within Glen Ogle in Perthshire, two of which obliterated sections of the road through the glen, trapping vehicles and necessitating the airlift of 57 people to safety. Most hillslope debris flows are initiated by shallow translational landslides, when the sliding mass of saturated soil and debris liquefies and begins to flow, though in gullies flood torrents may be transformed into debris flows by the addition of debris from the gully walls. Downslope flow of coarse debris takes the form of inter-clast collisions within the moving mass, which is partly buoyant in a mixture of water and mud.

**Figure 7.16** Debris flows on Sgùrr nan Fhir Duibhe (963 m), the easternmost summit of Beinn Eighe. Successive flows originating in upper slope gullies have formed tongues of quartzite debris over the darker Torridonian sandstone.

Gully exposures in debris cones and talus slopes often reveal stacked debris-flow deposits separated by buried soil or peat layers that represent periods of prolonged stability between flow events. Radiocarbon dating of these organic layers at various sites in the Highlands has shown that debris flows have occurred intermittently over at least the last 7000 years at such sites. However, the dated exposures represent only the most recent events, and it is likely that many Scottish debris cones began to accumulate immediately after deglaciation.

Some dating evidence also suggests that debris flows have become more frequent in Scotland within the past few centuries, possibly as a result of burning of heathland, deforestation or grazing pressure, all of which potentially increase the susceptibility of soil-mantled hillslopes to slope failure. There is also persuasive evidence that settlement expansion, woodland clearance and possibly grazing pressure in the Southern Uplands initiated gully erosion and enhanced debris-flow activity as early as the Late Bronze Age (~4.0 ka). At some sites in the Highlands where deep gullies have been eroded in thick drift deposits, debris flows occur every few years.

A particularly active site, visible from the A9 road, is the steep northeastern slope of An Torc in Drumochter Pass, where deep gullies have fed repeated debris flows onto large debris cones at the slope foot. Over the past 40 years there have been at least seven debris-flow events at this site, evident in the form of fresh levées and overlapping terminal lobes (Chapter 10).

**Conclusion**

The first edition of Archibald Geikie's classic book *The Scenery of Scotland* (1865) devoted over 100 pages to glaciation and its effects on the landscape, but the reader searches in vain for any mention of landslides. A century later Brian Sissons' equally classic *The Evolution of Scotland's Scenery* (1967) allocated just a single paragraph to the topic. As this chapter has shown, however, slope failures of various types and sizes are a pervasive characteristic of Scotland's mountain scenery; it is almost impossible to spend a day amongst the mountains without recognizing their effects, whether these take the form of small hollows marking the traces of shallow translational landslides, the tracks of debris flows, the sweep of rockfall talus slopes, a jumble of huge boulders marking a rock avalanche, or the ridge-crest scarps, bulging slopes and antiscarps that indicate the downslope shift of entire mountainsides.

Though our understanding of catastrophic rock-slope failures and rock-slope deformations is incomplete, it seems that the great majority of these large landslides represent paraglacial adjustment of mountain slopes, mainly within the five millennia that followed deglaciation, and that many catastrophic failures were triggered by earthquakes of a magnitude no longer experienced in Scotland. It would be premature to judge that large-scale catastrophic failures are extinct, however; even as you read these pages it is possible that a mountain slope somewhere in the Highlands is gradually creeping towards instability due to fracture of asperities and rock bridges along joints deep underground. A recent study based on the ages of catastrophic rock-slope failures in the Highlands concluded that the probability of large-scale catastrophic failure has not changed in the last 10,000 years. Sometime, somewhere, another will occur, though the likelihood is that it will be in a remote Highland glen, threatening only a few sheep and possibly a party of unfortunate hillwalkers.

The landslides that threaten infrastructure tend to be much smaller and are mainly due to translational sliding of soils or debris flows, which have buried or destroyed roads at the foot of steep soil-mantled hillslopes, as happened in Lochaber in 1953, the Cairngorms in 1956 and 1978, Glen Ogle in 2004 and recently (and with inconvenient frequency) the road between Arrochar and Inverary. Such events are sufficiently rare to be briefly newsworthy, but on the global scale of landslide disasters they are negligible.

From a geomorphological perspective, the most trenchant conclusion of recent research is the finding that many bedrock slopes and arêtes carry the scars of ancient landslides from which the runout debris was subsequently removed by glacier ice. Archetypal 'glacial' landforms such as corries and arêtes have evolved not simply through erosion by glacier ice during successive glacial stages, but also as a result of landslides and rockfall modifying mountain slopes during intervening interglacials. It is too much to claim, as a student of mine did, that glaciers 'just clear up the mess', but gravity has certainly played a much greater role in the evolution of Scotland's mountain scenery than has hitherto been appreciated.

# Chapter 8

# Aeolian landforms

## Introduction

As all visitors to Scotland's mountains are aware, strong, gusty winds frequently sweep across high ground. Mountain weather forecasts frequently warn of strong gusts and their consequences, in terms that range from cautionary ('may impede progress') to threatening ('severe buffeting') to downright intimidating ('progress impossible'). The frequency of strong winds on Scottish mountains reflects exposure to cyclonic systems (depressions) that track eastward across the North Atlantic Ocean. Strong winds generated by steep air pressure gradients within depressions are forced to rise over mountain barriers, accelerating airflow across summits, cols and plateaus, and generating turbulent eddies on lee slopes. Gale-force winds exceeding 80 km h⁻¹ (50 mph) are common on high ground. During the period 1884–1903, an average of 261 gales per year were recorded on the summit of Ben Nevis, and the *average* monthly wind speed recorded in recent years at the summit of Cairn Gorm (1245 m) has exceeded 56 km h⁻¹ (35 mph) during the winter months. Cairn Gorm summit also holds the record for the strongest gust ever recorded in the British Isles, of 277 km h⁻¹ (166 mph), and there is an unverified account of a gust of 301 km h⁻¹ (194 mph) – only slightly less than the strongest gust recorded anywhere on Earth.

Such powerful winds have produced some remarkable mountain landforms. Wind erosion has scoured plateau surfaces, creating sparsely vegetated expanses of boulders and gravel, as well as a range of smaller landforms. Most remarkably, particles deposited by wind have accumulated on some mountains to form thick accumulations of windblown sand. In this chapter we consider first the basic processes involved in aeolian (wind-related) erosion and deposition, then the effects of these processes on high ground in Scotland.

## Aeolian processes

Wind erosion involves two processes: *deflation*, which occurs when strong winds entrain exposed soil particles, and *wind abrasion*, which involves detachment of grains of rock or soil due to the impact of windblown particles near the ground surface. The size of particles entrained by wind generally increases with wind velocity. In wind tunnel experiments, sand grains 1 mm in size begin to move when wind velocities reach about 40 km h⁻¹ and particles 2 mm in diameter become mobile when wind velocity exceeds 50 km h⁻¹. In practice, though, stronger gusts are usually required to entrain soil particles, because of the effects of cohesion (wet particles sticking together) and soil compaction. Wind erosion is usually negligible where complete vegetation cover protects the underlying soil, but on unvegetated or partly vegetated plateaux in Scotland it has played a significant role in modifying plateau landscapes.

Grain size also determines the way in which particles are transported by wind. Clay- or silt-sized particles less than about 0.02 mm in diameter can be carried in suspension within turbulent airflows. Such fine particles may be lofted high into the atmosphere and can be transported long distances, just as far-travelled windblown dust from the Sahara Desert occasionally reaches the British Isles. Larger particles usually travel close to the ground. Sand particles up to about a millimetre in diameter mainly move in a series of steps by *saltation* or 'bouncing', usually no more than about 20–30 cm above the surface unless travelling over bedrock, frozen ground or ice. Impact by saltating grains may cause *reptation*, the downwind displacement of particles on the ground. Larger sand grains tend to move by rolling across the ground surface, a process referred to as *creep*.

Deposition of particles occurs when wind velocities fall below the threshold for particle movement. This often occurs where particles are blown onto more sheltered locations, such as lee slopes or the downwind sides of obstacles, or become trapped amongst vegetation. The nature of deposition depends on the mode of particle movement: creeping or saltating sand grains accumulate through *tractional deposition*, whereas accumulation of suspended silt particles is referred to as *airfall deposition*.

## Landforms produced by wind erosion
### Deflation surfaces

The most extensive products of wind erosion on Scottish mountains are *deflation surfaces*, broad expanses of bare ground where stripping of vegetation cover has exposed the soil to wind scour. On some plateaux this has winnowed away all soil particles up to 4–6 mm in size, leaving sterile surfaces that are armoured by boulders and carpeted with a gravel lag deposit. Deflation surfaces occur on plateaux underlain by all rock types in the Highlands but are rare in the Southern Uplands. They are particularly common on plateaux where the underlying rocks have weathered to produce sandy soil that is loosely packed and lacks cohesion, and hence is readily eroded by wind. Outstanding examples of deflation surfaces occur on the granite plateau of the Cairngorms and on the Torridonian sandstone mountains; the archetype is the broad northern plateau of An Teallach (Fig. 8.1a). Most deflation surfaces occur on plateaux and cols above 700 m, though they are present as low as 400 m on the windswept summits of Ward Hill on Orkney and Ronas Hill on Shetland.

There is evidence that some deflation surfaces originally supported soils and complete vegetation cover, but that this has subsequently been stripped away by wind erosion. In such locations, vegetated 'islands' of wind-blown sand and soil rise above the wind-scoured terrain (Fig. 8.1b). The margins of these remnants are currently eroding, suggesting that soil and vegetation cover were once much more extensive. Moreover, on sheltered slopes flanking some deflation surfaces there are thick accumulations of vegetation-covered aeolian sand deposits that represent the products of wind erosion of the adjacent deflation surfaces; good examples of paired

**Figure 8.1** (**a**) Deflation surface on the northern plateau of An Teallach in Wester Ross, showing the gravel lag that remains after finer soil bas been eroded by wind. (**b**) A remnant island of eroding sand rising above a deflation surface near the summit of Ward Hill, Orkney. The eroding margins of the sand suggest that sand and vegetation cover were formerly more extensive. (**c**) Deflation scars eroded into ericaceous vegetation cover at the summit of Monamenach (807 m) in the southeast Grampians. The scars are about a metre wide.

deflation surfaces and lee-slope sand sheets occur on the sandstone mountains of northwest Scotland, such as Slioch and Ben Mór Coigach. Luminescence dating of the timing of the most recent episode of sand accumulation on An Teallach has demonstrated that catastrophic erosion of soils on the northern plateau of that mountain commenced as recently AD 1550–1700; before then much of the plateau probably supported an extensive vegetation mat that protected the underlying soil from erosion. As outlined in more detail below, the timing of this event suggests that widespread stripping of soil and vegetation cover from the An Teallach plateau and from other summit plateaux in Scotland was triggered by the violent storms of the Little Ice Age of the 16th–19th centuries, a period of prolonged snow-lie and extremely stormy conditions.

### Wind-patterned ground

Not all wind-eroded plateaux are devoid of vegetation. Even the most sterile deflation surfaces support small clumps of grass, sedge or heather, and there is often a transition from completely vegetated plateau surfaces to completely denuded deflation surfaces. At one end of this spectrum is vegetation cover interrupted by *deflation scars*, small patches of lag gravels surrounded by eroding soil scarps up to about 0.4 m high (Fig. 8.1c). These develop where localized degradation of vegetation cover exposes soil that is loosened by frost heave and needle-ice growth, and subsequently eroded by the wind. More distinctive is the development of *wind-patterned ground* which takes the form of clumps, crescents or parallel stripes of vegetation on otherwise bare ground. *Wind crescents* are arcuate clumps of vegetation scattered across otherwise vegetation-free deflation surfaces, and *wind stripes* are straight or wavy lines of vegetation that alternate regularly with unvegetated wind-scoured ground (Fig. 8.2a).

The origin of these intriguing patterns is uncertain. On some plateaux, however, there is a transition from isolated deflation scars to linked deflation scars that grade into wind stripes and then, as the proportion of bare ground increases, into wind crescents and finally isolated clumps of vegetation. This transition suggests that wind-patterned ground has developed through progressive erosion of vegetation cover, with isolated deflation scars representing the first stages in vegetation stripping by frost and wind, and isolated clumps or crescents of vegetation representing advanced erosion of former vegetation cover.

**Figure 8.2 (a)** Wind stripes on Beinn a'Bha'ach Ard, Strathfarrar. The pole is about a metre long. **(b)** Turf-banked terraces on An Teallach, Wester Ross.

Studies of wind stripes have shown that these tend to be aligned at right angles to the dominant wind direction, and wind crescents tend to be concave upwind. Both types of feature are usually asymmetrical in cross-section, with a miniature eroded soil scarp and exposed roots on the upwind side and a gentler slope on the downwind side where ericaceous vegetation is progressively colonizing bare ground. Because erosion on the upwind side is counterbalanced by vegetation colonization on the downwind side, stripes and crescents migrate slowly downwind at rates of up to 20 mm per year. Studies of wind stripes in the Cairngorms have shown that the wavelength of individual stripe patterns (crest-to-crest distance) ranges from about 0.8 m to 1.4 m, and this is fairly typical of stripe patterns elsewhere on Scottish plateaux, although wavelengths of up to about 4 m have been observed. More puzzling is the surprisingly regular spacing of wind stripes and crescents at many locations

(Figure 8.2a). This may reflect a regular wavelength of near-surface air turbulence as strong winds sweep across stripe patterns, or is in some way regulated by the rate of vegetation colonization across unvegetated ground.

## Turf-banked terraces

Closely related to wind stripes are *turf-banked terraces*, which are step-like landforms with steep vegetated risers and gently sloping, sparsely vegetated treads or steps (Figure 8.2b). Such terraces are usually elongate across the slope, so they impart a 'staircase' appearance to upper slopes. Like other aeolian landforms, they are most common on rocks like granite and sandstone that have weathered to produce cohesionless sandy debris but occur on all lithologies, though they are absent where there is complete vegetation cover. Across northern Scotland the lowest altitude of terrace occurrence rises eastwards, from about 450 m on Rum and the Red Hills of Skye to 600 m on An Teallach and 700 m on Ben Wyvis. On the flanks of the Cairngorms they are limited to slopes above 800 m, but on the exposed hills of Orkney and Shetland they descend as low as 350 m. This pattern of altitudes emphasizes the role of exposure to westerly storm-force winds in terrace formation. They are absent from most of the mountains of the Southern Uplands but occur near the summit of Corserine (814 m) in the Galloway Hills.

Turf-banked terraces occur on a wide range of gradients (6–32°), though as the slope steepens the treads become narrower and the vegetated risers become steeper and higher. On Ronas Hill in Shetland, the bare treads of terraces are 0.8–4.2 m wide and up to 23 m long (across-slope), and the vegetated risers are 0.6–3.0 m wide. On An Teallach in NW Scotland, terraces on east-facing slopes have tread widths of 0.5–4.0 m and lengths of 4–25 m, with riser heights ranging from a few centimetres on gentle slopes to 1.4 m on the steepest slopes. Vegetation cover on risers tends to be dominated by grasses and heathers. On some rock types, particularly schists, broad turf-banked terraces with treads up to 12 m wide grade into active solifluction terraces, such as those described in Chapter 6.

The role of wind in terrace formation is also evident from their morphology. On sheltered east-facing slopes they are often horizontal across-slope, though individual terraces may be connected to their neighbours by an oblique vegetated riser. On other slopes, turf-banked terraces dip towards the direction of dominant wind. This relationship is beautifully illustrated on the

northern plateau of An Teallach (Fig. 8.3), where horizontal terraces on east-facing slopes are replaced by 'interconnected' terraces on northeast- and southeast-facing slopes, which in turn give way to westward-dipping terraces on north- and south-facing slopes. Turf-banked terraces tend to be absent or poorly developed on west-facing windward slopes.

The origin of turf-banked terraces remains contentious. An early commentator grasped the essence of terrace behaviour, describing the stony debris on terrace treads as 'a sea of moving material sweeping around islands of more or less stationary turf'. Subsequent research has vindicated this interpretation, showing that stones and soil migrate downslope across gentle terrace treads due to solifluction (particularly frost creep) at rates of a few millimetres per year, building up very slowly behind the vegetated riser. The initial formation of turf-banked terraces poses a greater problem. Some commentators have envisaged that such terraces form on slopes that originally supported complete vegetation cover, and represent expansion and interlinking of deflation scars to form wind stripes that have been modified into terraces by slow downslope creep of stony soil over bare ground. Others have suggested that selective vegetation colonization across initially bare slopes traps creeping debris and forms nascent vegetated risers that grow upwards and laterally as they trap downslope-creeping soil. A final possibility is that the risers form first on initially bare slopes as solifluction terraces that afford shelter to colonizing vegetation, which then anchors the riser and grows upwards to trap mobile debris at the lip of the tread. This chicken-or-egg problem has not been resolved, though observations of wind stripes on gentle slopes grading into turf-banked terraces as the slope steepens indicate that at least some terraces owe their origin to selective wind erosion of vegetated ground and subsequent modification of a stripe pattern by downslope creep of stony soil.

## Ventifacts

At a smaller scale, exposed boulder and bedrock surfaces on the higher parts of Scottish mountains sometimes exhibit evidence for abrasion by windblown sand or silt particles. Such abraded surfaces are termed *ventifacts* and are generally best developed within 30 cm of the ground surface (the maximum height reached by most saltating sand grains). Abrasion by sandblasting takes several forms, including faceted, indented, pitted or grooved rock or boulder surfaces (Fig. 8.4), but the most

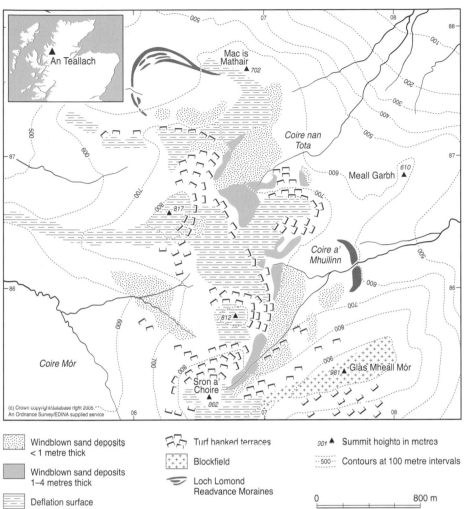

**Figure 8.3** Aeolian landforms on the northern plateau of An Teallach, Wester Ross. The windblown sand deposits on windward (west-facing) slopes are less than 0.5 m thick, but those on lee (east-facing) slopes are up to 4.0 m thick at the margin of the plateau deflation surface. Turf-banked terraces are horizontal on east-facing slopes but dip obliquely westwards on north- and south-facing slopes.

**Figure 8.4** Wind-abraded outcrop of Torridonian sandstone on An Teallach. Sandblasting has smoothed and pitted the rock, and eroded the base of the outcrop.

common effect on Scottish mountains is wind-polishing by impacting grains, creating a smooth, glossy surface on the upwind side of boulders or rock outcrops. The effects of aeolian abrasion are present on a wide range of rock types, including granite, gneiss, quartzite, sandstone and basalt, though they appear to be less common on schists. Many of the granite boulders that litter the Cairngorms exhibit the effects of wind abrasion, though the best-developed ventifacts in Scotland are probably wind-abraded boulders of fine-grained quartzite on plateaux in the Northern Highlands. The effects of recent wind abrasion are usually confined to west-facing boulders and rock outcrops, reflecting the present dominant wind direction, though some researchers have identified relict lichen-covered wind-abraded and wind-polished surfaces that may reflect changing wind directions under the harsh periglacial condition of the Lateglacial period.

## Aeolian deposits

Amongst the most intriguing landforms on Scottish mountains are gently undulating, vegetation-covered accumulations of windblown sediment that occupy plateaux, cols and upper slopes. Although early researchers described these as 'hill dunes', these deposits lack the structure and form of dunes and are more accurately referred to as *sand sheets* (Fig. 8.5). Sand sheets are particularly common on the Torridonian sandstone mountains of northwest Scotland, such as Ben Mór Coigach,

**Figure 8.5** (**a**) Sand sheet at the summit of The Storr in northern Skye. (**b**) Eroded upslope margin of the sand sheet at the head of Coire a'Mhuillin, An Teallach. (**c**) Pit excavated in the windblown deposits on the summit of The Storr. The pit is 2.9 m deep. (**d**) Section through the margin of the An Teallach sand sheets. The dashed line represents the boundary between the weathered lower-unit sand and the fresh upper-unit sand. The latter began to accumulate rapidly as a result of catastrophic wind erosion of the adjacent plateau within the period AD 1550–1700.

Cùl Mór and Slioch, but are also present on granitic rocks (as on the Cairngorms, the Red Hills of Skye and Ronas Hill in Shetland), on ultrabasic rocks on Rum, and on schists on mountains throughout the Highlands, though they are more common north of the Great Glen than in the Grampians. Fine-grained deposits of wind-blown silt and sand also mantle the upper slopes of some quartzite mountains in the far northwest, such as Foinaven and Arkle. The apparent absence of windblown sediments on many mountains may be misleading, as some mountain soils contain a component of airfall sediment. Where this dominates soil composition, the soils are termed *aeolisols*.

The presence of deposits of windblown sand on exposed summits that experience strong winds and occasional torrential rainstorms might seem bizarre, but investigation of a remarkable sand accumulation at the summit of The Storr (719 m) on the Trotternish Peninsula of Skye provides some answers. Visitors have commented on the lushness of the vegetation cover on the highest part of the mountain, little realizing that this consists of a cap of windblown sediment covering an area of 33,000 m² (Fig. 8.5a). These sediments are dominated by sand derived from the local basalts, but also contain silt and fine gravel, and even clasts up to 11 cm long. Investigation of this sand sheet has shown that it is over 2.5 m thick near the edge of the steep cliffs south of the summit (Fig. 8.5c) and thins westward and northwards away from the cliff edge. The size of the particles in the sand sheet shows a similar trend, progressively decreasing northwards and westwards away from the top of the cliffs.

These trends suggest that the windblown sediments capping The Storr were derived from the cliff below the summit. Wind tunnel experiments have shown that on meeting a vertical face airflow accelerates rapidly upwards, then decelerates as it passes over the top of the face. In a similar way, it appears that southerly gales meeting the cliffs of The Storr accelerate on meeting this barrier, entraining grains and small clasts that have been released from the cliff face by weathering. At the crest of the cliff, deceleration of airflow causes the suspended particles to rain onto the summit plateau, where they are trapped by vegetation and slowly accumulate, gradually thickening the sand sheet. Radiocarbon dating of organic material from near the base of the windblown deposits on The Storr has confirmed this interpretation. This shows that the sand began to accumulate sometime between 7.1 ka and 5.7 ka. The cliff that apparently formed the source of the sand was created by a major landslide (the Storr landslide) at ~6.1 ka. It therefore seems likely that exposure of the cliff by the landslide initiated accumulation of the summit sand sheet. Radiocarbon dating of organic material within the windblown sediment has also allowed calculation of average rates of sand accumulation, which range from about 14 mm per century near the edge of the sand sheet to roughly 60 mm per century in the thickest deposits near the crest of the southern cliffs. This gradual accretion of windblown sand probably continues at present.

The above findings demonstrate that some aeolian deposits mantling summits and plateaux in Scotland may be derived from adjacent cliffs, but where such cliffs are absent there is convincing evidence that sand sheets have formed through accumulation on lee slopes of particles blown from adjacent deflation surfaces. This has been demonstrated for the thick deposits of windblown sand that have accumulated downwind of the broad deflation surface on the northern plateau of An Teallach (Figs 8.1a and 8.3). On lee (east-facing) slopes these deposits reach a depth of up to four metres (Fig. 8.5b) and are composed primarily of sand with a small component of silt and occasional layers of fine gravel. Study of recent sand accumulation on lee slopes has shown that much of it initially accumulates on snowcover during winter storms, then is lowered onto the underlying vegetation as the snow melts; the vegetation then grows through the accumulating sand, securing it from erosion. On lee slopes the thickness of sand declines with distance from the edge of the deflation surface, demonstrating that the latter represents the source of the sand.

The An Teallach sand deposits comprise two distinct units: an upper unit of fresh, unweathered sand 1.0–2.2 m thick and a lower unit of weathered sand up to 2.4 m thick (Figure 8.5d). The two are locally separated by an unconformity that indicates erosion of the lower unit before deposition of the upper unit. Radiocarbon dating and pollen analyses of organic layers in the lower unit suggest that it began to accumulate by ~11.0 ka, when establishment of a heathland flora after disappearance of the last glaciers allowed trapping of windblown particles by vegetation. By ~8.0 ka, the deposits at the plateau edge had reached a thickness of 0.5 m, and slow accumulation continued until these were up to 2.4 m thick. There appears to have followed a long period of very limited sand accumulation, probably because a stable vegetation cover became established over much of the plateau source area.

The onset of upper-unit sand accumulation represents a catastrophic environmental change caused by widespread stripping of vegetation cover on the plateau and consequent erosion of underlying sandy soils. Luminescence ages obtained from the base of the upper sand unit place the onset of this event within the period AD 1550–1700. As noted earlier, this timing coincides with the Little Ice Age, a period of prolonged snow-lie (which may have degraded vegetation cover) and violent, gusty storms, which not only stripped the vegetation mat from the plateau, but also caused erosion of the underlying sandy soil and re-deposition of the eroded sand on adjacent slopes, allowing rapid accumulation of the upper sand unit. It is also possible that introduction of sheep to the mountain in the eighteenth or early nineteenth centuries may have triggered or exacerbated vegetation degradation; sheep are voracious grazers, cropping grasses to their roots and exposing the underlying soil. For sites near the plateau margin it has been calculated that whereas the lower sand unit accumulated at an average rate of 0.18–0.23 mm per year (similar to the rate of accumulation of the summit sand sheet on The Storr), the upper unit accumulated at an astonishing rate of 2.5–8.5 mm per year. It is fascinating to realize that the sterile deflation surface that now occupies the northern plateau of An Teallach probably supported a cover of grasses and sedges similar to that now occupying the summit of The Storr (Figure 8.5a) just a few centuries ago.

Similar catastrophic episodes of wind erosion are recorded in the aeolian sand deposits on other mountains in the form of a unit of fresh windblown sand overlying much older weathered sand on lee slopes. Pits excavated at the margins of sand sheets on Ben Mór Coigach near Ullapool and Carn nan Gobhar above Loch Mullardoch exhibit both units, implying widespread erosion of soils from adjacent plateaux over the past few centuries. The timing of the onset of such erosion has been established at three further sites by luminescence dating of the contact between the lower and upper sand units. On Ben Dearg Mhór (709 m) in

the Red Hills of Skye the upper sand unit began to accumulate sometime between AD 1500 and AD 1800, and on Fionn Bheinn (933 m) near Achnasheen sometime between AD 1500 and AD 1700. Both age ranges straddle those obtained for sites on An Teallach, suggesting that greatly accelerated wind erosion of plateaux was widespread during the Little Ice Age. Luminescence ages obtained for the base of the upper sand unit on A'Mharconaich (975 m) in the Drumochter Hills, however, place the onset of plateau erosion later, around AD 1900; here overgrazing rather than climatic stress may have been responsible for destabilizing the vegetation cover that protected soil from erosion.

## Conclusion

Wind action is often underestimated as a geomorphic agent. As this chapter has shown, however, soil erosion and deposition of sand by wind have played a distinctive role in the evolution of many Scottish mountain landscapes. An interesting implication is that the sparsely vegetated deflation surfaces that occupy many plateaux are not pristine landscapes that have remained unchanged since mountain summits emerged from the last ice sheet, as there is clear evidence from several sites that these sterile surfaces once supported widespread soil and vegetation cover, which has been extensively eroded by wind, particularly within the past few centuries. A further implication is that vegetation and soil cover on the higher parts of some Scottish mountains are more fragile and vulnerable to erosion than they might appear. The history of wind erosion on mountains such as An Teallach, Ben Mór Coigach and the Red Hills on Skye shows that opening of the vegetation cover by vegetation degradation, footpath erosion or overgrazing can potentially trigger expansion of deflation scars and progressive erosion of soil on windswept plateaux and cols. Predictions of climate change in Scotland emphasize a future increase in the frequency and ferocity of strong winds. Should these predictions prove accurate, catastrophic soil erosion may revisit some high plateau areas in the not-too-distant future.

# Chapter 9

# Fluvial landforms

## Introduction

Scotland's mountains are notoriously wet. Air masses and fronts moving across Scotland from the Atlantic Ocean are forced to rise sharply on meeting mountain barriers, causing the air to cool so that water vapour condenses and falls as rain or snow. As a result, total annual precipitation tends to increase with altitude. The mean annual precipitation recorded at 670 m altitude on An Teallach in Wester Ross, for example, is ~3500 mm, double that recorded at the foot of the mountain. All but the most easterly summits in Scotland receive on average over 1800 mm of precipitation per year, with some in the Western Highlands receiving over 4000 mm. Rainstorms with intensities exceeding 50 mm in 24 hours are fairly common, generating floods; these can be particularly dramatic if rain falls on snow cover, so that melting snow augments the volume of water reaching river channels. Moreover, where bedrock is at or near the ground surface, almost all stormwater rapidly reaches river channels, so that rivers have a flashy response to rainstorms, with runoff discharge rising rapidly to a flood peak. Such floods have washed away footbridges and even roads: a flash flood on the Allt Mor in 1978, for example, destroyed part of the approach road to the Cairngorm ski station. Where slopes are covered by thick drift deposits or peat, however, some rainfall is stored as groundwater, reducing flood peak runoff discharge during rainstorms. On the rare occasions when Scotland's mountains experience prolonged drought, it is flow of stored groundwater into river channels that maintains continuous runoff. At such times only the headwaters of mountain streams dry up completely.

## Mountain rivers

The mountain rivers of Scotland form a hierarchy of types that change in a downstream direction. *Headwater streams* occupy shallow channels high on mountainsides, and are often sourced from boggy cols or corries, or fed by groundwater that seeps through blockfields on upper slopes. On some mountainsides there is a conspicuous springline at the junction between permeable frost debris on upper slopes and relatively impermeable glacial drift deposits on lower slopes, with gullies terminating abruptly upslope along the springline (Fig. 5.17a). Headwater streams feed steep *mountain torrents* that course through miniature rock gorges or gullies and cascade over waterfalls before joining the rivers that drain tributary valleys, where two types of channel are often present: narrow *bedrock channels*, where rivers have cut through bedrock steps, and *alluvial channels*, where rivers flow over fluvial sands and gravels, usually in *meandering channels* that wander back and forth across the valley floor. Such tributary channels eventually join *trunk rivers* that flow along the floors of glacial troughs.

On the floors of broad tributary valleys and glacial troughs, most rivers take the form of *wandering gravel-bed rivers*, comprising gently sloping alluvial reaches alternating with those where the channel is incised in bedrock and descends in a series of waterfalls and pools. On the alluvial reaches, rivers commonly follow a single meandering channel, with eroding banks where flow is rapid on the outside of meander bends and deposition of gravel in the form of *point bars* on the inside of meander bends (Fig. 9.1). Over time, bank erosion causes meandering channels to migrate slowly downstream and to extend laterally across valley floors, becoming progressively 'loopier' until the neck of the meander is breached during floods, leaving an abandoned meander on the floodplain. Many wandering gravel-bed rivers also have *braided* reaches occupied by two or more channels separated by gravel bars (Fig. 9.2). These can be unstable, and subject to channel switching (*avulsion*) during extreme floods. So although the bedrock reaches of river valleys in the Scottish mountains are fixed in location, over decadal timescales the position of channels on alluvial reaches may change drastically through meander migration

**Figure 9.1** Meandering channel of a wandering gravel-bed river, Langadale, North Harris, showing bank erosion on the outside and downstream banks of meanders, and point bar deposition on the insides of meander bends.

**Figure 9.2** Braided reach of the River Feshie, Cairngorm Mountains. The main channel occasionally shifts across the floodplain during extreme floods. Low Holocene alluvial terraces flank the present floodplain. © Lorne Gill/SNH.

and cutoff, or through avulsion. The River Etive in the western Grampians, for example, beautifully illustrates such alternation of stable bedrock channels and unstable alluvial reaches. The River Feshie and River Dee on the margins of the Cairngorms are amongst the most unstable rivers in Scotland, exhibiting channel switching, gravel-bar migration and increased braiding in response to extreme flood events, then a tendency to reoccupy a single dominant channel during periods that are uninterrupted by major floods. Studies of channel change based on historical maps have shown that the main channels of the River Tay and its tributary the Tummel have migrated locally up to a kilometre across their floodplains since 1783, though flood embankment construction now limits the freedom of these and other major upland rivers to reoccupy much of their floodplains.

The floodplains occupied by wandering rivers often appear featureless, but from above abandoned channels are evident; often these are occupied by marshes, and less frequently by shallow *oxbow lakes*, where water is ponded in abandoned meander channels. Occasionally channels are bordered by embankments called *levées*, sometimes of natural origin, though those along major rivers have usually been constructed to prevent flooding, and by splays of sediment, where rivers have burst their banks during floods. Where they enter lochs or the sea, Scottish rivers deposit sediment to form *deltas* (Fig. 9.3). The River Etive, for example, once entered Loch Etive about a kilometre farther inland than now, but since the retreat of the last glacier in Glen Etive the river has infilled the head of the loch with sediment, thereby extending its floodplain seaward.

## River terraces

Bordering the present floodplains of many rivers in trunk and tributary valleys are low-gradient terraces that descend gradually downvalley (Fig. 9.4). These often extend continuously along valley margins on both sides of the present floodplain, but in steep tributary valleys terraces are often represented only by isolated fragments. The presence of river terraces implies that rivers have initially deposited a thick sediment cover on valley floors, then cut down into these deposits, leaving former floodplain fragments high and dry above the present active floodplain. Such behaviour can be explained in two ways.

**Figure 9.3** The River Coe, a typical wandering gravel-bed river, where it enters Loch Achtriochtan. The river occupied the abandoned channel nearest the road before AD 1875. The main channel then switched to the far side of the floodplain, before migrating to its present position. Where the river enters the loch a small delta has formed; the loch is gradually being infilled with fluvial sediment.

**Figure 9.4** River terraces in Glen Garry, central Grampians. The three higher terraces are outwash terraces, formed when the river cut down into glacifluvial deposits. The low terrace adjacent to the river may represent floodplain aggradation then incision during the Holocene.

First, there may have been a change in *base level*, which controls the local level of a river. For the lowest reaches of rivers that terminate at the coast, sea level controls the base level. If sea level rises, rivers tend to deposit sediment near their mouths, but if sea level falls, rivers cut down into their floodplains in response, and remnants of the former floodplain form terraces. In upland areas, however, individual alluvial reaches are graded to local base levels represented by bedrock steps. If a river succeeds in cutting down through a bedrock step, lowering the local base level, then it will also incise the floodplain immediately upstream, again leaving remnants of the former, higher floodplain as river terraces.

The second way in which terraces form depends on the balance between sediment supply and the discharge regime of a river. If the volume of sediment supplied from upstream exceeds the capacity of a river to transport all sediment, then net sediment deposition occurs on the valley floor and the floodplain undergoes *aggradation* (thickening). Conversely, if the capacity of a river to transport sediment exceeds the sediment supply from upstream, the river will tend to cannibalize sediment

from its floodplain, cutting down to a lower level and leaving remnants of the former higher floodplain as river terraces. Almost all river terraces in Scotland fall into four categories: *outwash terraces* and *kame terraces* that were deposited by rivers draining former glaciers, *alluvial terraces* that formed later, after the disappearance of the last glaciers, and *strath terraces*, which differ from others in that they have formed through fluvial incision of bedrock.

## Outwash terraces and kame terraces

The great majority of terraces flanking the present floodplains of major rivers in Scotland are outwash terraces that represent abandoned and incised proglacial floodplains or sandar (Chapter 5). During the final periods of glacier retreat, braided rivers draining the fronts of retreating ice margins deposited thick accumulations of sand and gravel that are now represented by flights of outwash terraces up to 40 m above present floodplains (Fig. 9.4). Such terraces support a microtopography of abandoned braided channels, and some are pitted with kettle holes. The most impressive outwash terraces occur outside the limits of the Loch Lomond Readvance glaciers, for example in the Angus glens of the SE Grampians, in the Findhorn Valley and in Glen Feshie. In all of these locations terraces occur at several levels, with the uppermost and oldest terraces representing the original sandar surfaces, and lower terraces representing subsequent floodplain levels. If the terraces are 'paired' at similar heights on either side of a valley, then pulsed river incision is implied, with stable floodplains forming as an equilibrium was reached between sediment supply and discharge regime, followed by downcutting of the river to a new level as sediment supply from upstream diminished, or the discharge regime of the river changed to increase its capacity for sediment removal. Where opposing terraces are 'unpaired', then continuous incision by a migrating river is more likely, with the river eroding sediment from first one side of its floodplain then, at a lower level, from the opposite side, then sweeping back to erode its previous floodplain leaving a terrace fragment. Few outwash terraces in Scotland have been dated. High terraces more than ~4 m above the present floodplain probably represent initial floodplain aggradation and subsequent floodplain incision and terrace formation during the Lateglacial period (~17.0–11.7 ka), though continued river incision and terracing of outwash deposits may have persisted into the Holocene.

As outlined in Chapter 5, many of the highest terraces on valley-side slopes in the Highlands and Southern Uplands are kame terraces formed by deposition of sediment by rivers flowing between the margins of former glaciers and the adjacent valley-side slope. As these glaciers thinned, kame terraces were left perched high above present floodplains. Although in some locations it is difficult to distinguish between kame terraces and outwash terraces, the former are generally less continuous, often present on only one side of a valley, and frequently associated with valley-side meltwater channels. Terraces more than about 40 m above present floodplains are almost certainly kame terraces. Good examples occur along the northern flanks of the Cairngorms and along the northern shore of lower Loch Etive (Fig. 5.20).

### Alluvial terraces

Alluvial terraces have formed where postglacial rivers first deposited sediments, leading to floodplain accretion, then incised these floodplains leaving low terraces above the present channel. In the Scottish Highlands such low terraces are widespread on the margins of alluvial reaches in most trunk and major tributary valleys, but rarely rise more than 4 m above the present floodplain. The ages of such terraces have been determined by radiocarbon dating of organic material exposed in eroded riverbanks or the base of peat overlying terrace fragments, and though only a handful of such dates exist, all suggest that floodplain accretion was succeeded by incision and terrace formation within the last 5000 years. For example, the low terrace bordering Edendon Water, which drains the east Drumochter Hills, represents the level of the floodplain about 2500 years ago, after which the river has cut down about two metres to its present level. The cause of this widespread change from net floodplain accretion to fluvial incision and terrace formation in the Highlands is unknown, but similar low alluvial terraces in the Southern Uplands have been attributed to Neolithic or Iron Age human impacts, particularly burning and clearance of woodland. Such activities are thought to have released sediment into river systems, causing floodplain aggradation on valley floors, followed by river incision and terrace formation as sediment input subsequently diminished.

### Bedrock channels and strath terraces

In locations where upland rivers flow over bedrock, they cascade over successive bedrock steps called *knickpoints*, forming low waterfalls. Detachment of bedrock by floodwaters armed with cobbles and boulders causes knickpoints to retreat upstream, leaving bedrock terraces (*strath terraces*) on either side of narrow gorges. Various studies have shown that the amount of knickpoint retreat since deglaciation is related to catchment area and thus flood discharge: for small streams, it amounts to a few tens of metres, but on larger rivers knickpoints have retreated several kilometres. TCN dating of strath terraces abandoned through a combination of knickpoint retreat and channel incision around Loch Linnhe in the western Grampians suggests that knickpoint retreat following deglaciation at the end of the Loch Lomond Stade (~11.7 ka) was initially rapid, then declined, a pattern that has been attributed to a progressive reduction in the supply of cobbles and boulders that form the 'tools' that enable rivers in spate to detach bedrock at knickpoints. Although ancestral bedrock channels and gorges were probably initiated through erosion by glacial meltwater rivers, the present dimensions of such channels have been determined by postglacial knickpoint retreat and river incision, though these processes now operate only very gradually.

## Alluvial fans

At locations where steep, fast-flowing mountain torrents meet the broad, gently sloping floors of glacial troughs, reduction in the velocity of the torrent results in deposition of pebbles, cobbles and boulders. Over time, this debris builds up to form an alluvial fan, so-named from the tendency of such sediment accumulations to splay out across valley floors in a fan shape. The steepest alluvial fans, with gradients up to about 12° at the apex, occur at the mouths of steep tributary valleys or gullies (Fig. 9.5), and such upland fans often support bouldery lobes and levées deposited in part by torrential floods and sometimes by confined debris flows (Chapter 7). More extensive, low-gradient alluvial fans also occur at the confluences of major rivers, such as where the River Feshie meets the River Spey near Kincraig, and where the River Quoich joins the River Dee near Braemar.

The largest alluvial fans in Scotland are either *outwash fans* deposited by glacial meltwater streams, or relict paraglacial landforms of Lateglacial age, formed as rivers entrained glacial deposits then deposited these as sands and gravels along valley sides or at river confluences. Some, such as those in Glen Banchor, exceed a kilometre in width at their lower ends. Such fans often terminate on high outwash terraces, such as the Allt Fhearnagan fan, south of Achlean in Glen Feshie. However, subsequent floodplain incision and abandonment of high terraces

has resulted in trenching of the original fans by their parent streams, and formation of lower fans that are nested within the initial landform (Fig. 9.5).

The most spectacular Lateglacial fans in Scotland are those in Glen Roy, in the Lochaber district of the western Grampians. As outlined in Chapter 5, Glen Roy was occupied by a sequence of ice-dammed lakes during the Loch Lomond Stade, and supports large relict alluvial fans, in particular three known as the Brunachan, Reinich, and Canal Burn fans (Fig. 9.6). Although these were originally interpreted as paraglacial fans that accumulated prior to drowning of Glen Roy under the sequence of ice-dammed lakes, recent research suggests that they accumulated first in shallow water then in deeper water as lake levels rose, and therefore represent lake deltas overlain by river gravels deposited as lake levels fell. Floodplain incision after final lake drainage has resulted in fluvial incision of these fans, producing inset terraces and frontal bluffs. A fourth large fan located where Glen Turret meets Glen Roy also deserves mention, not least because its origin has provoked more acrimonious debate than probably any other Scottish landform. Some researchers have interpreted this impressive feature as a paraglacial fan formed prior to the Loch Lomond Stade; others (more plausibly) as an ice-contact outwash fan that accumulated at the snout of a glacier that occupied Glen Turret during the Loch Lomond Stade.

**Figure 9.5** Two generations of alluvial fans in An Caorann Mór, near Loch Cluanie. The main fan is a paraglacial landform that formed soon after deglaciation (~11.7 ka) through reworking of hillslope drift deposits by debris flows and mountain torrents. The smaller inset fan (lower right) represents re-deposition of sediment eroded from the main fan.

**Figure 9.6** The Canal Burn fan, upper Glen Roy, most of which was deposited in a former ice-dammed lake. Floodplain incision by the River Roy caused the Canal Burn to entrench the fan. The terraces bordering the River Roy were formed as it cut down into its floodplain.

On all the large fans described above, sediment accumulation was succeeded by stream incision before the end of the Lateglacial period at ~11.7 ka. A later generation of small alluvial fans, however, has formed during the Holocene at locations where steep tributary valleys or gully systems meet low-gradient valley floors. These small Holocene fans represent episodic erosion of drift-mantled slopes during extreme rainstorm events, entrainment of the resulting sediment by mountain torrents or debris flows, and deposition of sand and gravel on valley floors and low alluvial terraces. Researchers have been able to reconstruct the histories of some of these small fans through radiocarbon dating of buried peat layers at sites where the fan architecture has been exposed by later erosion. In this way it has been shown that the small alluvial fans in Scottish mountain valleys did not accumulate incrementally, but through a series of major depositional events separated by long periods of quiescence. A small fan in the Edendon Valley of the Drumochter Hills, for example, accumulated during just three rainstorm-generated depositional events, dated to ~2.2 ka, 1.9 ka and 0.8 ka; the entire fan probably formed in just a few hours over a timescale of two millennia.

A remarkable feature of small alluvial fans for which radiocarbon dates have been obtained is that all appear to have been deposited during the late Holocene (after ~4.2 ka). Researchers studying small fans in the Southern Uplands and northwest England have attributed this late onset of fan accumulation to periods of settlement expansion and upland land use change, particularly through woodland clearance, burning and grazing pressure. Such changes are thought to have lowered the threshold of stability of drift-mantled slopes, rendering them more susceptible to gully erosion and associated deposition of fan sediments during extreme rainstorms. In the Southern Uplands, sporadic gullying events started during the Late Bronze to Iron Age (~4.0–2.0 ka), with particularly intensive periods of gully erosion and fan accumulation at AD 700–900, AD 1100–1300 and AD 1450–1550. Whether similar land-use pressures initiated late Holocene erosion and concomitant fan deposition in the Highlands has not been established, though pollen assemblages and charcoal recovered from organic material underlying a small alluvial fan deposited about 600 years ago in Glen Etive suggest that woodland burning and resulting upslope erosion were responsible for fan accumulation. Pollen assemblages in the buried peat horizons of the Edendon fan, however, produced no evidence for burning or other changes in vegetation cover prior to the three runoff events that built the fan.

## Conclusion

The highest river terraces in Scotland represent outwash deposits that accumulated during retreat of the last ice sheet, when a combination of powerful meltwater streams and abundant sediment supply resulted in the infill of glacial troughs and tributary valleys with thick accumulations of sands and gravels. Most large alluvial fans in Scotland were also deposited in the wake of the retreating ice sheet. Since then, floodplain and fan incision have predominated, forming flights of terraces along troughs and lower nested fans within their Lateglacial ancestors. There is evidence in some locations, however, for renewed floodplain aggradation and fan deposition during the late Holocene, some of which may have been triggered by release of sediment into rivers as a result of hillslope gullying following deforestation, which in the Southern Uplands at least appears to be related to settlement expansion and woodland clearance. Our understanding of Lateglacial and Holocene alluvial changes in the Highlands, however, is underpinned by just a handful of radiocarbon ages from widely separated sites, making it difficult to discern any consistent pattern of river behaviour. This is a research field that remains largely untilled.

## Postscript: into the Anthropocene

In Chapter 4 we noted that many scientists believe that we have recently entered a new geological epoch, the *Anthropocene*, a view that acknowledges the role of human activity in influencing environmental change (notably climate change and its multiple effects, such as shrinking glaciers, thawing permafrost and sea-level rise) and certain geological processes, such as increased soil erosion and sediment load transported by rivers. There is limited agreement as to when the Anthropocene began. Human activity has certainly affected ecosystems for at least 40 ka, as demonstrated by the fact that extinctions of large ice-age mammals such as mammoths and mastodons (and even other hominid species, such as the Neanderthals) closely followed migration of modern humans to hitherto unsettled parts of the globe. It is, however, difficult to pinpoint a period or event when human activity began to constitute a dominant ecological or geomorphological process. Some scientists have advocated the beginnings of agriculture at 15–12 ka as initiating the Anthropocene; others place the invention of the steam engine and beginning of the European industrial revolution around AD 1780 as the key development; others still have narrowed the commencement of the Anthropocene to 16 July 1945, when the first atomic device was detonated in the New Mexico desert.

The Scottish mountains have been described as eternal, but as we have seen, this is the case only on a human timescale. Over long geological timescales Scottish mountains and volcanoes have risen and been eroded to their roots. Over the Quaternary timescale of 2.6 million years, glaciers and ice sheets have waxed and waned, sea levels have risen and fallen, and permafrost has aggraded and decayed. Over the 14,000–17,000 years since ice-sheet deglaciation, frost and wind have modified upper slopes and plateaux, mountainsides have crumbled, and rivers have deposited then eroded floodplains. Nothing is static, nothing endures and nothing is eternal, even though within our brief individual lifetimes this may appear to be so. As James Hutton remarked over two centuries ago, there is indeed no vestige of a beginning, nor any prospect of an end.

This book has described the processes that have affected Scotland's mountain landscapes since the last glaciers disappeared, but some of the most obvious recent changes to mountain scenery are anthropogenic, such as estate roads bulldozed across the southeast Grampians, proliferating wind turbines and micro-hydroelectric schemes, and the unsightly scars of clear-felled forestry. Furthermore, predictions of climate change suggest that over the next century Scotland will experience more frequent extreme storm events that may affect mountain landscapes through increased occurrence of shallow landslides and debris flows on drift-mantled slopes, and enhanced erosion of fragile mountain soils by slopewash and wind erosion. Such changes are already underway: annual precipitation totals for Scotland exhibit a general (though oscillatory) rise over the past century, with unprecedented high rainfall totals and annual river discharges over the past few decades, and a well-documented increase in the frequency of flood events.

Perhaps the most pervasive recent impact on the higher parts of Scotland's mountains, though, has been the emergence of unplanned footpaths as legions of hillwalkers wend their way to the summits along the shortest route from the nearest car park. Footpath erosion is a geomorphic process that involves a sequence of vegetation degradation, exposure of bare soil, then progressive deepening and widening of the pathway by trampling and erosion of soil by wind and runoff. Studies of footpath erosion on Scottish plateaux have shown that some combinations of soil and vegetation cover are fairly robust under trampling pressure but others are subject to rapid

**Figure 9.7** Footpath erosion on granite-derived soil in the Mamores. Erosion is extensive on moss-heath communities dominated by *Racomitrium lanuginosum* (foreground) but contained within stable narrow pathways on grass-heath communities where *Nardus stricta* and *Carex bigelowii* dominate (background).

degradation (Fig. 9.7). On slopes, however, unplanned footpaths often deteriorate rapidly into shallow stream-filled, boulder-floored gullies. Valiant efforts to repair footpaths on popular routes using stepped slabs and drainage diversion have sometimes been undermined (literally) by walkers descending along footpath margins, triggering renewed erosion. Perhaps we have yet to learn from the nineteenth-century estate owners who constructed robust, well-graded, well-drained stalkers' paths that have survived over a century of use by the simple expedient of using a gradual zig-zag line of ascent rather than a staircase to the summit.

In Scotland we have inherited some of the most diverse mountain landscapes on Earth, the products of a long and tumultuous geological history, moulded to their present form by glaciers, landslides, rivers, frost and wind. We also enjoy the freedom to explore our mountain heritage, to wander where we will across summits, plateaux and ridges, to camp in lonely glens or watch glowing sunsets redden the high snowfields. With these glorious privileges, however, comes a responsibility not merely to cherish Scotland's mountain landscapes, but also to preserve them for those who will follow our long-vanished footprints in the snow.

# Chapter 10

# Key sites

## Introduction

The sites described below have been selected to illustrate the diversity of Scotland's mountain landscapes. Many others would serve equally well. Every journey through or over Scottish mountains, however, encounters some of the landforms described in previous chapters. The barren ice-moulded gneiss hills of the Outer Hebrides, the sandstone inselbergs of Sutherland, the high tableland of Gaick, the quartzite ridges of the Grey Corries and the rolling summits of the Southern Uplands are left for the reader to explore. Armed with imagination, you can fill the glens with glaciers, envisage the collapse of mountainsides, and witness the changes wrought on mountain landscapes by frost, wind, rockfall and floods.

National grid references and GPS co-ordinates are given below where necessary to allow location of key sites and their component landforms.

## An Teallach, Wester Ross
### (NH 069843; 57.8065, -05.2517)
### Introduction

From the east, the view of An Teallach is striking: a pinnacled mass that glows orange in the early morning sunlight or is silhouetted black against the sunset. The mountain culminates in the twin peaks of Bidein a'Ghlas Thuill (1062 m) and Sgùrr Fiona (1059 m) and is composed of reddish Torridonian sandstones that dip gently southeastwards, though its three eastern summits are capped by Cambrian quartzite outliers that represent the continuation of a quartzite escarpment east of the mountain. The Torridonian rocks are of Neoproterozoic age (~1000–800 Ma) but remarkably unaltered, preserving the bedding of the original fluvial deposits. The Cambrian rocks were deposited much later (~530 Ma) and take the form of a pebbly conglomerate (Fig. 2.4d) overlain by fine-grained quartzite and 'pipe rock', a white quartzite riddled with grey mineralized worm casts.

An Teallach represented a formidable obstacle in the path of the last ice sheet, which flowed northwestwards across the area, though striae show that the base of the ice was diverted around the massif (Fig. 10.1). Erratics of schist and sandstone occur at nearly 900 m on the southernmost quartzite summit (Sàil Liath) and must have been carried upwards within the ice sheet before being deposited amid blockfield debris. Similarly, a conspicuous belt of quartzite erratics on the northern plateau (Fig. 5.11) represents uphill transport of debris derived from the escarpment east of the mountain. End and lateral moraines in the Ardessie valley west of the massif were deposited during the Wester Ross Readvance of ~15.3 ka, implying prior emergence of the An Teallach summits as the last ice sheet thinned.

### Loch Lomond Readvance moraines

During the Loch Lomond Stade of ~12.9–11.7 ka six glaciers reoccupied the corries of An Teallach and deposited lateral and end moraines (Figs 7.13b and 10.1). The most interesting lateral moraines are those fronting the two eastern corries, Glas Tholl and Coire Toll an Lochain. The moraines emanating from below the north-facing cliffs in these corries are much larger than those originating below the gentler south-facing slopes. This moraine asymmetry reflects debris supply: below the north-facing cliffs, the glaciers advancing out of the corries evacuated rockfall and rockslide debris that had accumulated at the foot of the cliffs before the readvance and deposited this reworked debris in large moraines; rockfall directly onto the glacier surfaces also contributed to building large moraines. From the gentler south-facing slopes there was no equivalent debris supply, and consequently the moraines are much smaller. This contrast in moraine size beautifully illustrates the synergistic relationship between glacial erosion and landslides, demonstrating the importance of rockfall and rock-slope failures in widening corries, and the role of glacier ice in evacuating the resultant debris.

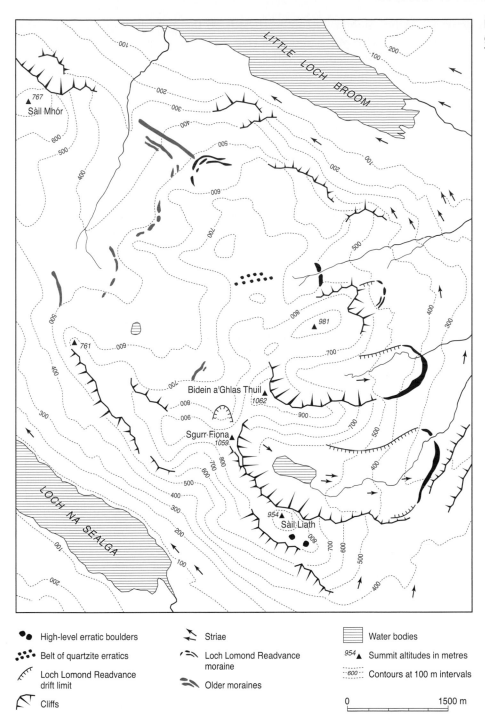

**Figure 10.1** Glacial geomorphology of An Teallach.

| | |
|---|---|
| •● High-level erratic boulders | |
| ⋮⋰⋱ Belt of quartzite erratics | |
| ⌒ Loch Lomond Readvance drift limit | |
| ⌒ Cliffs | |

↙ Striae

⌒ Loch Lomond Readvance moraine

〰 Older moraines

▭ Water bodies

954▲ Summit altitudes in metres

⋯600⋯ Contours at 100 m intervals

0             1500 m

This synergy is also illustrated by the pinnacled southeast ridge between Sàil Liath and Sgùrr Fiona. The northern wall of the arête takes the form of steep slabs that represent the failure planes of former rockslides (Fig. 7.12), but the volume of talus at the foot of these cliffs is very limited, implying removal of landslide runout debris by former glaciers.

## Aeolian landforms

The undulating, sparsely vegetated northern plateau of An Teallach represents a deflation surface that is developed on a blockfield of boulders embedded in sand (Fig. 8.3). Like other blockfields in Scotland, that on the northern plateau was preserved under cold-based ice during the last ice-sheet glaciation. This is demonstrated by a belt of quartzite erratics, derived from the Cambrian quartzite escarpment to the east, which crosses the plateau near the head of Coire à Mhuillin and was deposited on the blockfield surface as the last ice sheet thinned (Fig. 5.11).

The role of wind erosion in modifying the blockfield is evident in removal of all fine particles from the surface, leaving a carpet of fine gravel amidst the boulders (Fig. 8.1a). The eroded material has accumulated as wind-blown sand deposits up to 4 m thick on sheltered slopes northwest of the plateau (Fig. 8.5b). Much of the sand initially accumulated on or within snow cover, then was trapped by vegetation as the snow melted. Sand deposition began in the early Holocene, but an upper unit of fresh, unweathered sand up to 2.2 m thick (Fig. 8.5d) provides evidence of widespread stripping of soil and vegetation from the plateau after AD 1550–1700. This catastrophic event coincides with the colder, stormier conditions of the 'Little Ice Age' of the 16th–19th centuries and was probably caused by vegetation degradation under late-lying snow and consequent erosion of exposed soil by strong winds.

At the eastern margins of the northern plateau, flights of turf-banked terraces have developed where debris creeping downslope by solifluction has been arrested by bands of vegetation that form steep steps or risers, creating a 'staircase' of unvegetated treads and vegetated risers (Fig. 8.2b). On gentle north- and south-facing slopes similar terraces are aligned obliquely across the slope, invariably dipping westward, into the direction of dominant winds. These features probably originated as wind stripes that have been modified by downslope creep of debris to form horizontal or oblique terraces.

## Torridon, Wester Ross
### (NG 958569; 57.5559, -05.4149)
### Introduction

Glen Torridon rivals Glen Coe as a candidate for the most spectacular glacial trough in Scotland. The glen is flanked to the north by the massive bulk of Liathach, a pinnacled arête culminating in the conical peaks of Mullach an Rathain (1023 m) and Spidean a'Choire Liath (1054 m), and its scree-covered neighbour, Beinn Eighe, which reaches its highest point on Ruadh-stac Mór (1010 m). Glen Torridon and much of the surrounding high ground is underlain by Torridonian sandstone, but the eastern part of Beinn Eighe and several of the south Torridon summits are capped by white Cambrian quartzites. The unconformity that separates these two rock types dips

southeastwards, so the quartzite cap thickens in that direction (Fig. 2.10). Successive glaciers and ice sheets moved westwards across the area, eroding a breach that extends from Kinlochewe to Loch Torridon, and another through the gap between Liathach and its northern neighbours (Fig. 5.1b). During the Loch Lomond Stade, the Torridon area was occupied by an icefield that terminated near the head of Loch Torridon and near the southern end of Loch Maree. At this time glacier ice occupied most of the Torridon glens, encircling or partly encircling the mountains. Both Beinn Dearg (914 m) and Maol Chean-dearg (933 m) were nunataks at this time, islands of rock rising above the ice.

### The Torridon corries

The Torridon mountains host some of the deepest, largest and most spectacular corries in the British Isles. The north face of Liathach, for example, is indented by six corries, with precipitous sandstone headwalls, sidewalls and buttresses that rise up to 500 m above corrie floors. Where corries have developed in Cambrian quartzites, however, they assume a different form, with cliffs up to about 100 m high surmounting steep scree-covered slopes. This difference reflects different responses to glacial and periglacial processes. The depth of the corries on the Torridonian rocks suggests that despite their intrinsic strength and stability they have been vulnerable to deepening by erosion under successive glaciers and ice sheets. The well-jointed Cambrian quartzites, however, have succumbed to freeze–thaw weathering under the harsh periglacial conditions of the Lateglacial period, forming the mantle of bouldery scree that skirts corrie walls. Queen of the Torridon corries is Coire Mhic Fhearchair on the northwestern flank of Beinn Eighe (NG 945604; 57.5847, -05.4395). Here a thin outcrop of quartzite caps the spectacular 300 m high sandstone cliffs and gullies of the Triple Buttress, one of the most popular sites in Scotland amongst rock climbers.

### Loch Lomond Readvance moraines

An exceptional variety of Loch Lomond Readvance moraines occurs in Torridon. At the mouth of Glen Torridon, just east of the road, twin end-moraine ridges mark the limit of the Loch Lomond Readvance. A beautiful lateral moraine ascends for over 3 km across the western slopes of Liathach from near the coast of Upper Loch Torridon (NG 875574; 57.5566, -05.5538) to 490 m altitude and represents the limit of a former outlet glacier from Coire Mhic Nobuil that terminated just below present sea level. Across the same valley, a rare example of a medial moraine ridge descends southwards from 450 m (NG 881605; 57.5847, -05.5465) to 350 m and marks the former confluence of ice flowing southeast from Beinn Alligin with ice flowing south from Coire Mhic Nobuil. North of Liathach, recessional moraines mark the oscillatory eastward retreat of the former ice margin (Fig. 5.16), and on the heather-covered slopes south of Loch Gaineamhach (NG 831660; 57.6316, -05.6348) is a remarkable site where bouldery lateral moraines deposited by a Loch Lomond Readvance outlet glacier truncate (almost at right angles) lateral moraines deposited during the Wester Ross Readvance some 3000 years earlier.

By far the most eye-catching and photogenic moraines, however, are the hummocky moraines of Coire a'Cheud-cnoic (literally 'the Valley of a Hundred Hills', though one researcher counted over 300 individual hummocks). From the nearby car park (NG 958569; 57.5559, -05.4149) these appear as a chaotic assemblage of mounds and conical hummocks (Fig. 10.2) which has been variously interpreted as representing *in situ* stagnation moraines, push moraines or thrust moraines. The chaotic appearance of these moraines from the car park is deceptive, however. Viewed from above they can be seen to comprise two elements: broad, nested recessional moraines, aligned oblique to the valley axis, and, superimposed on these, longitudinal ridges aligned south–north towards Glen Torridon. The former are thought to represent recessional moraines formed at the margin of a glacier that retreated southwards during the closing stages of ice-sheet deglaciation, probably within

the period ~15–14 ka. Superimposed on these older moraines are elongate glacial bedforms (fluted moraines) formed by streamlining of till under the glacier that moved northwards to join the Glen Torridon glacier during the Loch Lomond Stade.

## Landslides

Although large postglacial rock-slope failures are comparatively rare on Torridonian sandstone and Cambrian quartzite, three spectacular catastrophic landslides are located in the Torridon area. Around 4400 years ago, release of nine million tons of rock from the south-facing cliff below the summit of Beinn Alligin (Fig. 7.1) produced the largest rock avalanche in Scotland. Such was the energy of this cascading mass of rock that huge boulders surged upslope onto the opposite valley side and formed a tongue of debris that travelled over a kilometre along the valley floor. The cause of this landslide is uncertain, but the failure scar is bounded on both sides by upwards-converging fault scarps, suggesting that tremors caused by activation of one or both faults could have triggered cliff collapse.

A similar though smaller failure involving about 0.6 million tons of rock occurred around 15.0 ka on the northwest face (NG 845676; 57.6466, -05.6129) of Baosbheinn. Here the mass of descending rock rebounded on the level ground at the slope foot to form an arcuate boulder ridge up to 31 m high that is draped over a lateral moraine deposited by the Wester Ross Readvance at ~15.3 ka. The third of this trio of great landslides occurred at ~12.4 ka when the quartzite rockwall forming the east face of Maol Chean-dearg (NG 930496; 57.4892, -05.4554) collapsed. The runout debris forms a broad lobe of quartzite boulders that extends over 500 m over the valley floor, burying the underlying sandstone bedrock and terminating abruptly at marginal ridges. This unusually long runout suggests that the landslide occurred onto a residual Loch Lomond Readvance glacier, which carried the boulders to their present position.

**Figure 10.2** Coire a'Cheud-cnoic, the 'Valley of 100 Hills' in Glen Torridon. These apparently chaotic hummocks represent overprinting of recessional moraines by later fluted moraines.

## The Trotternish escarpment, Isle of Skye
(NG 455627; 57.5824, -06.2593)
### Introduction
The Trotternish eescarpment extends for 23 km along the spine of the Trotternish Peninsula in northern Skye and is composed of a thickness of over 300 m of

stacked westward-dipping basalt lava flows (Fig. 2.4c) that represent the earliest phase (~58 Ma) of Palaeogene igneous activity on Skye. These lavas overlie shales, into which have been intruded thick westward-dipping dolerite sills. The escarpment consists of cliffs up to 250 m high and culminates in The Storr (719 m) and eight other peaks over 500 m. During the last glacial maximum, ice from the Cuillin Hills flowed northwards across Trotternish, feeding a major ice stream that drained the last ice sheet ice towards the edge of the continental shelf via the North Minch. Most of Trotternish was deglaciated around 16.6 ka, though a small glacier reoccupied Coire Cuithir (NG 470590; 57.5501, -06.2305) during the Loch Lomond Stade.

## The Trotternish landslides

The Trotternish landslides are the most famous in Scotland. Displaced rock masses extend along the foot of the escarpment, where there is an inner region of tabular blocks, ridges and pinnacles, and an outer zone of hummocky landslide terrain (Fig. 10.3). The former represents landslides that have occurred since deglaciation, and the latter comprises much older landslide blocks that have been over-ridden by successive ice sheets. Resting against the scarp are huge displaced blocks, such as Dùn Dubh (NG 441666; 57.6165, -06.2868), and farther out from the scarp are massive, isolated tabular blocks such as Leac nan Fionn (NG 453704; 57.6513, -06.2708) and Cleat (NG 447669; 57.6916, -06.2771). These detached

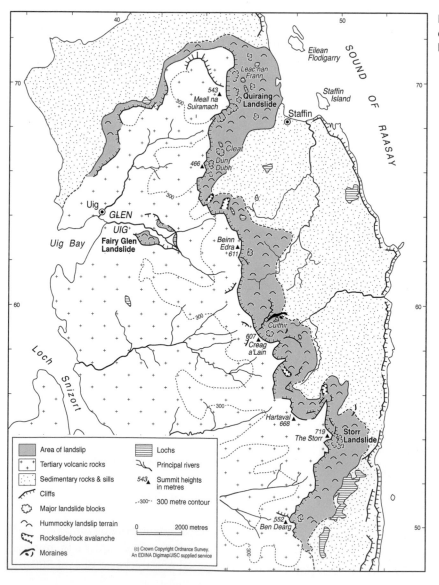

**Figure 10.3** The Trotternish escarpment, showing the extent of landslide terrain.

blocks appear to represent horizontal movement of intact lava blocks over the underlying shales, a phenomenon unique to Trotternish.

The most spectacular landslides on Trotternish are the Storr and Quiraing landslides. The entire south face of The Storr collapsed at ~6.1 ka to create a large hollow, Coire Faoin (NG 497537; 57.5041; -06.1800) bounded by basalt cliffs 200 m high. The undercliff zone is a labyrinth of lava blocks, narrow clefts and pinnacles of shattered rock, of which the 49 m high Old Man of Storr is the most impressive (Fig. 10.4). These pinnacles are tilted eastwards away from the cliffs, suggesting that the lava pile foundered on the underlying shale and fragmented as it did so. This 'topographie anarchique' (as described by a French geomorphologist) is the most photographed landform in Scotland and receives over 70,000 visitors each year.

The haunting architectural weirdness of the Storr landslide is rivalled by that of its northern neighbour, the Quiraing (NG 453692; 57.6405, -06.2695). This is the largest landslide complex in Scotland, covering an area of 8.5 km², and extends 2.2 km from the escarpment to the coast. As with the Storr landslide, the inner zone of the Quiraing comprises an array of detached blocks, corridors, ridges and pinnacles with evocative names like 'the Needle' and 'the Prison' (Fig. 10.5), and an outer zone of tilted and ice-moulded landslide blocks. The structure of this huge slope failure is complex. The largest intact landslide block, Leac nan Fionn, has moved laterally about 250 m from the backscarp without tilting and with negligible vertical displacement. Other blocks are back-tilted, implying deep-seated rotational landsliding. Although most of the Quiraing complex is stable, coastal erosion of the toe of the landslide still triggers intermittent seaward movement. The Geological Survey Memoir notes that 'the main road near Flodigarry is frequently dislocated', and there are reports of tremors as this sleeping giant stirs.

So impressive are the Storr and Quiraing landslides that they have overshadowed a remarkable landslide in Glen Uig, the Fairy Glen landslide. This occupies the south side of the glen (NG 415631; 57.5837, -06.3265)

**Figure 10.4** The Storr landslide, which occurred about 6100 years ago and involved eastwards sliding and tilting of blocks of lava that simultaneously fractured to form pinnacles such as the Old Man of Storr.

**Figure 10.5** The Prison, a detached block of the Quiraing landslide, formed by rotational sliding of the lava pile over underlying shales.

and extends 370 m from the low backscarp. Most of the slide consists of rounded lava blocks, but standing proud of these is an isolated basalt tower, Castle Ewen, that has moved over 150 m horizontally away from the backscarp. One possible explanation for this extraordinary behaviour is that ice-rich permafrost may have formed in the underlying shale during the Lateglacial period, so that subsequent thaw of ice within the shale generated high water pressures that permitted Castle Ewen to glide gracefully to its present position.

### Relict talus accumulations

Along parts of the escarpment that have not experienced postglacial landsliding, sweeping aprons of vegetation-covered talus have accumulated as a result of intermittent rockfall from the cliffs above. These talus accumulations are locally incised by deep gullies, allowing researchers to investigate the structure of the talus and reconstruct the history of talus accumulation. Radiocarbon dating of buried soil layers in the talus

suggests that almost all rockfall accumulation occurred during the Lateglacial period prior to 11.7 ka, probably in response to a combination of paraglacial stress release and frost weathering of cliff faces. Studies of the talus stratigraphy have shown that they have experienced extensive modification by debris flows that have redistributed sediment downslope.

### Aeolian and periglacial landforms on The Storr

Under the grasses and sedges at the summit of The Storr lies a vegetation-covered cap of windblown sediments that covers an area of 33,000 m$^2$ (Fig. 8.5a). This remarkable deposit began to accumulate around 6.1 ka when the present cliffs were exposed by the Storr landslide. Since then, gusty winds have blown grains weathered from the cliff face upwards over the summit, where they have been trapped by vegetation cover. The deposit thins away from the crest of the cliffs, from a maximum depth of 2.9 m near the summit (Fig. 8.5c) to less than 0.4 m around its margins. Miniature frost-sorted patterned ground occupies areas of wind-scoured bare ground (deflation scars), and earth hummocks occur on the aeolian sediment cover. On adjacent slopes, active solifluction is evident in the form of turf-banked terraces and small vegetated solifluction lobes that have advanced downslope to bury Holocene peat deposits. Recent movement of ploughing boulders is also evident from upslope furrows up to 4 m long and gaps between mobile boulders and the soil upslope. A small basalt tor rises above a deflation surface about 500 m north of the summit.

## The Cuillin Hills, Isle of Skye
(NG 440224; 57.2205, -06.2424)
### Introduction
Visitors to Skye cannot fail to be impressed by the striking contrast between the rugged serrated battlements of the Cuillin Hills and the lower domes of their neighbours, the Red Hills (Fig. 3.9). Both represent the uplifted roots of Palaeocene volcanoes that once towered 2000 m or so above the adjacent terrain and are now exposed by removal of the overlying rocks. Their contrasting topography reflects their contrasting geology: the tough, massive gabbros of the Cuillin Hills have evolved by rockslides, rockfalls and the cumulative effects of corrie formation, whereas the joined granites of the Red Hills have weathered more readily into rounded hills, skirted by screes and locally scalloped by corries. Both have been intruded by numerous basalt dykes, but whereas those in the Cuillin have eroded preferentially to create chasms and chimneys (Fig. 2.4b), those in the Red Hills have little topographic expression.

Erratics carried westward from the adjacent mainland are abundant along both the northern and southern coasts of Skye but absent from the Cuillin and Red Hills, demonstrating that during the last ice-sheet glaciation these mountains supported an independent ice dome that covered all summits and diverted mainland ice northwards into the Minch and southwards towards the Sea of the Hebrides. As the encircling ice sheet retreated around 16.0 ka, ice nourished from the mountains of Skye expanded across the vacated terrain, depositing a moraine on the neighbouring island of Soay at ~15.1 ka. Glaciers vanished from Skye under the warmer conditions of the Lateglacial Interstade (~14.7–12.9 ka), but during the ensuing Loch Lomond Stade renewed expansion of glacier ice resulted in the build-up of an icefield 156 km² in area amid the mountains of Skye, flanked by smaller glaciers that reoccupied the corries of the Cuillin

and eastern Red Hills (Fig. 4.15). These glaciers reached their maximum extent at ~12.4 ka and probably disappeared by ~11.5 ka in response to the rapidly warming climate of the early Holocene.

### The Cuillin ridge
The Cuillin ridge forms a 12 km long arc of rocky summits that extends from Gars-bheinn (895 m) in the south to Sgùrr nan Gillean (964 m) in the north and constitutes the most relentless and exposed traverse of mountains in the British Isles (Figs. 1.1b and 3.9a). It is the finest arête in Scotland, sculpted not only by glacier ice (as commonly supposed) but mainly by rockfalls and rockslides that have exposed steep slabs, pinnacles and vertiginous rockwalls along its entire length. Evidence for the passage of glacier ice across the ridge is limited to outcrops of glacially moulded gabbro and perched glacially deposited boulders below 860 m, mainly in the cols that separate the southern summits between Gars-Bheinn (895 m) and Sgùrr Alasdair (992 m). The predominant role of rockfall in fashioning the ridge is evident in the great sweeping taluses of rockfall boulders that skirt the foot of the neighbouring cliffs (Fig. 7.13a). These represent the rockfall debris deposited since the retreat of the last glaciers, but similar rockslides and rockfalls must have occurred during previous interglacials, progressively reducing the Cuillin ridge to its present form.

Fortunately for those who brave the airy ridges and pinnacles of the Cuillin ridge, the coarse-grained gabbro bedrock offers the most tenacious boot-to-rock contact of any rock in Scotland. This is because preferential weathering and detachment of biotite crystals have created exceptionally rough rock surfaces dominated by protruding crystals of pyroxene and feldspar. This advantage is absent on fine-grained gabbro, however – most notoriously on that of the Inaccessible Pinnacle (986 m), the most difficult ascent of all the Munros (Fig. 2.5). Interestingly, a slab of rock seems poised to collapse from the east side of the 'In Pinn', potentially making it even more 'inaccessible' than is now the case.

### The Cuillin corries
The Cuillin ridge is flanked by corries that face all aspects, reflecting its role as a source of glaciers that radiated away from the ridge during successive glaciations. In size and grandeur these rival the Torridon corries, being deeply indented by precipitous cliffs flanked by talus slopes of bouldery rockfall debris (Fig. 5.6c). During the Loch

Lomond Stade, the corries west of the ridge were occupied by small glaciers, 0.7–4.6 km² in area, whereas those east of the ridge fed the Cuillin Icefield (Fig. 4.15). The limits of the glaciers that occupied the western corries are defined by drift limits and end moraines, the finest of which forms an arc of boulder ridges that descend from 300 m to 140 m (NG 443182; 57.1831, -06.2331) on the low ground fronting Coir a'Ghrunnda. This moraine enjoys the distinction of being the first to appear on a map, in a paper published by the pioneering Scottish glaciologist James Forbes in 1846. The floors of most corries exhibit outstanding examples of glacial scouring and moulding of bedrock, with striae, grooves, gouges and chattermarks being preserved on the surfaces of many outcrops. Ice moulding reaches its apogee on the rock outcrops that dam the lochan in Coire Làgan (NG 443208; 57.2064, -06.2358) where streamlined gabbro outcrops represent what are probably the most beautiful whalebacks in Scotland (Fig. 5.8c).

# Glen Roy, Lochaber
**(NN 297853; 56.9276, -04.7999)**
## Introduction
Glen Roy is probably the most-visited 'Ice Age' site in Scotland not only on account of the remarkable preservation of the shorelines of lakes dammed by glacier ice during the Loch Lomond Stade (Fig. 10.6), but also because of its historical significance for the development of the Glacial Theory (Chapter 5). The glen is underlain by Dalradian metasedimentary rocks, and probably lay close to the main north–south axis of the last ice sheet. The timing of ice-sheet deglaciation in this area probably occurred in the interval ~15–14 ka.

## Lake shorelines
The three main lake shorelines at 260 m, 325 m and 350 m are represented by horizontal benches ('parallel roads') with average widths of 10–11 m. These have formed mainly by wave erosion of hillslopes at the margins of former glacier-dammed lakes, though frost action under the severe periglacial conditions of the Loch Lomond Stade may also have contributed to shoreline development by detaching rock debris that was removed on ice floes. The widest shorelines occur at the head of the glen, where fetch and therefore wave energy were greatest. As outlined in Chapter 5 (Fig. 5.22), the 260 m shoreline formed initially as a glacier advancing east from the Great Glen dammed a large lake in Glen Roy and Glen Spean that drained eastwards over a bedrock threshold at 260 m near Kinloch Laggan. Further advance of the ice margin severed the Glen Roy

**Figure 10.6** The lake shorelines ('parallel roads') of Glen Roy at 260 m, 325 m and 350 m altitude (photograph by David Evans).

lake from that in Glen Spean, forcing the former to drain first over a col at 325 m, then, when that was blocked by ice advance, across the 350 m col separating Glen Roy from the upper Spey Valley. The sequence of lake levels was reversed during ice retreat, and ultimate drainage of the Roy–Spean lake took the form of a catastrophic *jökulhlaup* when over 5 km³ of water escaped along the Great Glen.

The shorelines are best viewed from a car park and viewpoint in the lower glen (NN 297853; 56.9276, -04.7999), from where a path worn by generations of enthusiastic geography students leads to the 260 m shoreline. The widest and most impressive shorelines, however, are those at the head of the glen (NN 352929; 56.9978; -04.7146), which reach 30 m in width and exhibit steepening of the backslope by wave erosion (Fig. 5.21).

### The Glen Roy fans

Flanking the floor of Glen Roy are eight large fans, all of which have now been entrenched and terraced by their parent streams. The most impressive of these are the Brunachan (NN 318896; 56.9669, -047683), Reinich (NN 330908; 56.9781, -04.7494) and Canal Burn (NN 357920; 56.9899, -04.7058) fans (Fig. 9.6). These are composed of gravels overlain in turn by lake sediments then a final layer of gravel, indicating that they represent deltas deposited in the former ice-dammed lakes. The origin of another large fan, the Turret Fan (NN 342922; 56.9911, -04.7306) is contentious, but most authorities consider that this impressive landform is an outwash fan deposited in front of a former glacier in Glen Turret. After the lakes drained, the rivers that deposited the fans have cut down over 20 m, leaving terraces inset into the fans.

### Landslides

The most striking of the rock-slope failures in Glen Roy is that which destroyed the lake shorelines opposite Brae Roy lodge (NN 342914; 56.9840, -04.7301). Less conspicuous from the glen is a large rock-slope deformation below the summit ridge of Beinn Iaruinn (800 m), as this landslide is limited to upper slopes and did not affect the shorelines farther downslope. The occurrence of these landslides has been linked to evidence for earthquake activity in the form of dislocation of three segments of the lake shorelines, which demonstrates movements along faults since shoreline formation. It is uncertain whether fault activation and consequent

earthquake activity is attributable to rapid unloading of the valley as the lake drained or (as elsewhere in Scotland) to differential glacio-isostatic uplift.

## Glen Coe, Western Grampians
### (NN 171569; 56.6680, -04.9863)
### Introduction

Glen Coe has acquired notoriety as the site of the massacre in 1692 of members of the MacDonald Clan by soldiers under the command of Archibald Campbell, Earl of Argyll. Under leaden winter skies it is indeed a sombre, brooding place, flanked to the south by the great buttresses (truncated spurs) of the Three Sisters, which form outliers of Bidean nam Bian (1150 m; Fig. 10.7), and to the north by the steep, rocky slopes of the Aonach Eagach, which culminates in Sgòrr nam Fiannaidh (967 m). Geologically, the mountains of Glen Coe represent a site of cauldron subsidence, where the sinking of fault-bounded blocks into an underlying magma chamber triggered a suite of explosive eruptions at ~421 Ma as the rising magma met surface water. The result was the accumulation of a thick pile of andesitic and rhyolitic ignimbrite sheets, lavas and sills, fine examples of which can be seen in the exposed faces of the Three Sisters.

### Glaciation

Glen Coe is a trough breach where, during successive glaciations, ice from the Rannoch Moor ice centre escaped westward through the glen, gradually eroding the preglacial north–south drainage divide and truncating the north faces of the Three Sisters. Erratics of Rannoch Moor granite are strewn along the floor of the glen as far as the shores of Loch Linnhe, and numerous striae record former westward ice flow. During the Loch Lomond Stade, ice nourished in the corries south of the glen joined an outlet glacier flowing west from Rannoch Moor, filling the valley with ice up to an altitude of 650 m at its eastern end and 450 m near its western exit, where the Coe glacier was confluent with a glacier that advanced down Loch Linnhe to terminate near Kentallen. The maximum altitude achieved by the Coe glacier at this time is recorded by trimlines cut intermittently across the southern slopes of the Aonach Eagach, notably at ~500 m on the southern slopes of Sgòrr nam Fiannaidh (NN 133574; 56.6710, -05.0486), which shows that the ice covering the site of the Clachaig Inn about 12,000 years ago was at least 400 m thick. In

**Figure 10.7** The Three Sisters of Glen Coe: stacked sills, lavas and ignimbrite sheets that form truncated spurs (photograph by John Gordon)

the highest part of the glen, the River Coe flows through a fault-guided slot (NN 176566; 56.6655, -04.9780) that was incised by a glacial meltwater stream.

## Postglacial landforms

The lower slopes of Glen Coe above the A82 road are crossed by some of the largest debris cones in Scotland. These represent episodic deposition of bouldery detritus by debris flows and mountain torrents emanating from deep bedrock gullies (eroded dykes) on the southern flanks of the Aonach Eagach. Lobes and splays of fresh boulders deposited during recent rainstorms are evident on some cones, and on several occasions the road has been over-run by debris-flow deposits, necessitating installation of larger culverts and the excavation of trenches across fans to act as sediment traps. Studies of the sizes of lichens on the largest cone (NN 156573; 56.6710, -05.0111) indicate that successive debris flows have covered the entire cone surface within the past 300 years, though the history of debris accumulation on this and other Glen Coe debris cones probably spans the entire Holocene.

The lower reaches of the River Coe downvalley from Achtriochtan Farm (Fig. 9.3) are typical of the wandering gravel-bed rivers that occupy the floors of glacial troughs in the Highlands. Study of historical maps has shown that during the past two centuries the main

channel of the river has migrated southwards from one side of its floodplain to the other, probably through avulsion (rapid channel switching) during an extreme flood event, before returning to its present position. Loch Achtriochtan itself is slowly shrinking due to sediment deposition and the progradation of the delta built by the river where it enters the loch.

The most spectacular postglacial landforms in the Glen Coe area are those in in Coire Gabhail, a hanging valley that drains northwards into the glen. The mouth of the valley has been dammed by two steep cones of large landslide boulders (NN 166557; 56.6710, -04.9936) that represent separate catastrophic failures of the ignimbrite cliffs upslope (Fig. 10.8). The source of the larger (southwestern) cone is a scar roughly 150 m wide below the summit of Gearr Aonach (692 m) and the smaller cone was sourced from a funnel-shaped scar about 100 m farther northwest. The larger cone was deposited by a landslide about 1700 years ago, and involved collapse and disintegration of roughly 0.6 million tons of rock. The smaller cone overlaps the larger, implying that the associated landslide occurred more recently, and the freshness of the debris cover suggests that it may be no more than a few centuries old.

Following blockage of the valley by landslide debris, coarse alluvial sediments have accumulated upvalley from the landslide, forming a gently sloping braided

**Figure 10.8** Coire Gabhail, where two cones of landslide debris have trapped alluvial deposits upvalley, forming a braided floodplain that lacks a river (photograph by John Gordon).

floodplain (Fig. 10.8). This floodplain is unique in Scotland in that it now lacks a river, except after exceptionally prolonged, intense rainstorms. This is because the streams that drain into the valley head sink into the floodplain gravels, and there is no surface flow in the valley until springs emerge at the northern end of the landslide debris. Coire Gabhail is popularly known as 'The Lost Valley': a silent glen, isolated from the bustle of Glen Coe by the landslide debris that blocks its exit.

## The Cairngorms
### (NJ 005040; 57.1161, -03.6447)
### Introduction

The Cairngorm massif incorporates an unrivalled variety of landforms that have developed over a range of timescales: a preglacial upland granite landscape modified by selective glacial erosion and various postglacial processes. The Cairngorm granite represents a pluton intruded around 425 Ma into Dalradian metasedimentary rocks during the later stages of the Caledonian Orogeny. This granite mass was probably unroofed by ~400–380 Ma as a result of vigorous erosion of the overlying Dalradian rocks and was uplifted to roughly its present altitude during the Cenozoic era. An important feature of the Cairngorm granite is the presence of *linear alteration zones*, belts of hydrothermally weathered granite up to 200 m wide that represent zones where heated groundwater circulated through fractured rock, causing chemical weathering. Linear alteration zones were exploited by river erosion to form the preglacial valley system, much of which was subsequently modified by selective glacial erosion to form deep troughs incised into the Cairngorms plateaux.

Schist erratics carried by the last ice sheet from sources west of the Cairngorm massif are abundant along its northern and southern margins but absent from the higher parts of the massif. This distribution tells us that during the Last Glacial Maximum the Cairngorms supported a persistent independent ice divide, a dome of ice that rose above, diverted and merged with the encircling ice sheet. TCN dating of boulders on moraines indicates that the last ice sheet withdrew from the flanks of the Cairngorms at 16.5–15.5 ka, though residual glaciers probably persisted later in high corries. During the Loch Lomond Stade only small glaciers formed in the Cairngorms, most of which were confined to corries.

## Palaeosurfaces, blockfields and tors

Two extensive palaeosurfaces are present on and around the massif (Fig. 3.12). The higher, the Cairngorm summit surface, is represented by an undulating plateau of wide, shallow valleys and cols surmounted by domed summits such as Ben Macdui (1309 m), Braeriach (1296 m), Cairn Gorm (1245 m) and Beinn a' Bhuird (1196 m). A scarp ~200 m high separates this surface from the Glen Avon embayment, a plateau at 700–800 m altitude that covers ~65 km², mainly in the northeast of the massif, and represents part of the extensive Eastern Grampian surface, the palaeosurface that dominates much of the Grampians from the Spey southeastwards to the Highland boundary. Both palaeosurfaces represent etchplains that are believed to have developed to roughly their present form through pulsed uplift, slope retreat, fluvial erosion and deep weathering during the Cenozoic era. Exposures of sandy gruss up to 10 m deep in some headwater valleys on the Cairngorm summit surface, for example in Coire Raibeirt (NJ 004049; 57.1242; -03.6437) and south of the summit of Ben Avon (NJ 132014; 57.0954, -034342), provide evidence of the role of chemical weathering in the evolution of the plateau topography.

The granite tors of the Cairngorms represent, according to one commentator, 'the best example of a glaciated torfield in the world'. They occur over an altitudinal range of 600–1240 m, are up to 24 m high and range in form from small rounded bedrock protrusions to towers the size of a house; the largest occupy areas of over 2000 m² (Fig. 6.4a). The tors represent areas of massive granite surrounded by terrain where the granite is more densely jointed. As a result, prolonged differential weathering and erosion have reduced the level of the surrounding plateau surface, leaving the tors upstanding. TCN dating of the Cairngorm tors suggests that the oldest may have emerged more than half a million years ago through lowering of the surrounding surface at rates of up to ~30 mm ka⁻¹, equivalent to 30 m of lowering in the last million years. This indicates that the present palaeosurfaces are not pristine Cenozoic surfaces inherited from preglacial times, but have been lowered by weathering and erosion, though the general relief of the palaeosurface has been preserved.

Surrounding the tors are extensive granite blockfields, usually less than a metre thick, that have formed through frost weathering of bedrock under periglacial conditions. In some areas these comprise openwork boulders, but most consist of boulders embedded in coarse sand derived from granular weathering of granite. Most exposed boulder surfaces have been rounded in this way, but their undersides are usually angular, implying that granular weathering has affected only exposed surfaces. Like the tors, the Cairngorm blockfields survived burial under glacier ice at the Last Glacial Maximum (and indeed earlier glacial episodes) because the ice cover on high ground was frozen to the underlying substrate and capable of only very limited erosion. Many tors have nonetheless experienced modification by glacier ice, which has removed large blocks of the original tors, locally depositing these a short distance away. In extreme cases, tors have been reduced to their foundations, forming tor plinths such as that near Cairn Gorm summit (Fig. 6.4b).

## Landforms of glacial erosion

The Cairngorms represent *par excellence* a landscape of selective linear erosion. Whereas successive ice caps and ice sheets that occupied the high ground were cold-ice-based with limited erosive potential, glaciers occupying the intervening and surrounding valleys were warm-ice-based and capable of sliding over and eroding the underlying terrain, progressively widening and deepening these valleys to form Icelandic-type glacial troughs with deep rock basins at their heads (Gleann Einich and Glen Avon) and the glacial breaches of the Lairig Ghru and Lairig an Laoigh. Over 30 large arcuate corries indent Cairngorm summit surface, and high-level col breaches have also contributed to segmentation of plateau areas. Several corries are clustered to form compound corries, such as those south of the ski station, and the corries overlooking Loch Einich. Most corrie floors are over 900 m above sea level, and thus not only much higher than corrie floors in the western Highlands and Hebrides, but also higher than many Scottish mountain summits. Only a handful of Cairngorm corries support flooded rock basins produced by overdeepening of corrie floors by subglacial erosion, but almost all are skirted with talus or mantled by scree indicative of corrie widening by weathering and rockfall since deglaciation.

The Cairngorms therefore represent a landscape of paradox. The plateau areas are modified preglacial palaeo-surfaces, studded with tors that have emerged through differential weathering and erosion of the surrounding ground, and mantled by blockfields that probably date back to the last interglacial or earlier. This well-preserved ancient landscape has been dissected and fragmented by vigorous glacial erosion operating along or at the heads of the original fluvial valleys to form deeply entrenched troughs, glacial breaches and arcuate corries. Ascent of the Cairngorms thus involves a journey from a landscape dominated by glacial erosion to an undulating plateau only slightly modified by the ravages of glacier ice.

## The Loch Lomond Readvance

During the Loch Lomond Stade of ~12.9–11.7 ka, 16 or 17 small glaciers formed in some of the corries fringing the plateau. The extent of many of these is defined by spreads of boulders that terminate abruptly on corrie floors, though end moraines also define former glacier limits in a few locations, notably on the lip of Coire an Lochain north of Braeriach (NH 943006; 57.0842, -03.7456) and on the floor of Coire an t-Sneachda (Fig. 10.9). Hummocky recessional moraines in the southern part of the Lairig Ghru have been interpreted as indicating that two glaciers nourished in the corries to the west advanced into the glen, and it is likely that the larger of the two (in Glen Geusachan) may also have been fed by a small plateau ice cap. The average equilibrium line altitude of the Loch Lomond Readvance glaciers in the Cairngorms was about 900 m, much higher than that of glaciers in the western Highlands, such as those of the Mull and Skye icefields (250 m and 277 m respectively). The limited extent and high altitude of the Cairngorm glaciers is explained by scavenging of snow from Atlantic airmasses by the large icefield that developed to the west (Fig. 4.14), so that only light snowfall occurred on the Cairngorms to feed glacier growth.

**Figure 10.9** The footprint of a small glacier that occupied Coire an t-Sneachda in the Cairngorms during the Loch Lomond Stade. The former extent of the glacier is marked by an end moraine and, on the slope opposite, a line of large boulders. Several recessional moraines occur within the glacier limit.

## Postglacial landforms

Although large rock-slope failures are much less common on granite terrain than on schists, rockwall collapse at a few sites in the Cairngorms has spread aprons of large boulders across adjacent valley floors. In upper Strath Nethy a cover of boulders from two or more rock-slides extends almost continuously for two kilometres along the west side of the valley, burying the underlying till deposits. A similar drape of rockslide boulders occurs opposite the site of the former Sinclair Hut in the Lairig Ghru (Fig. 10.10), and less extensive deposits of coarse bouldery rockslide runout debris occur in Coire Beanaidh (NH 951014; 57.0916, -03.7328) and below the cliffs of Carn Etchachan (NJ 006008; 57.0874, -03.6418). The ages of these landslides have been established by TCN exposure dating: the Strath Nethy and Lairig Ghru landslides occurred very soon after ice-sheet deglaciation, and the Coire Beanaidh and Carn Etchachan slides occurred at ~14.3 ka and ~13.6 ka respectively. All were apparently caused by opening up of steep, slope-parallel dilation joints (sheeting joints) as the source rockwalls responded to unloading from under the cover of the last ice sheet, with consequent release of large rock slabs that broke up during movement and on impact with the slope foot.

Intermittent small-scale rockfalls have formed extensive talus accumulations below trough walls and corrie cliffs. Most are relict landforms that accumulated mainly within a few millennia following deglaciation, and many are completely vegetated and incised by gullies that show that erosion has replaced rockfall accumulation as the dominant geomorphic activity on talus slopes. Those that flank both sides of the Lairig Ghru, for example,

are extensively gullied, and partly buried by the tracks of numerous debris flows that occurred during intense rainstorms in 1956 and 1978. Debris flows are not the only agent responsible for degrading the talus slopes of the Lairig Ghru: several avalanche boulder tongues occur in the valley, at sites where snow avalanches have stripped debris from the upper parts of talus slopes and deposited it as stony lobes or tongues on the valley floor. The finest examples are those that dam the Pools of Dee (NH 973007; 57.0858, -03.6962) near the highest part of the pass, where powerful avalanches have transported clasts across the valley and a short distance up the opposite slope, forming boulder tongues up to 7 m thick.

On upper slopes, the most impressive and widespread postglacial landforms are bouldery stone-banked terraces (on gently sloping ground near slope crests) and stone-banked lobes, both with frontal risers up to about 3 m high (Fig. 6.5d). These features are thought to have formed through gradual movement of a layer of frost-heaved boulders that were rafted downslope by solifluction of the underlying soil under the severe periglacial conditions of the Lateglacial period when permafrost was present.

The effects of wind erosion on the Cairngorm plateau are evident in extensive deflation surfaces, occasional areas of wind stripes, and polished or pitted boulders or rock outcrops that have been sandblasted to form ventifacts. Occasional low vegetation-protected islands of sandy soil represent remnants of formerly more extensive soil and vegetation cover, though there is limited evidence for the catastrophic wind erosion that affected some plateaux in the Northern Highlands, as aeolian sand deposits are thin or absent on most lee slopes.

**Figure 10.10** Hummocky landslide debris draped over thick till deposits at the northern end of the Lairig Ghru.

## Drumochter Pass, Central Grampians
### (NN 629770; 56.8641, -04.2505)
### Introduction

The A9 road that links northern Scotland and the central belt reaches its highest point (462 m) in Drumochter Pass. West of this bleak valley lie the West Drumochter Hills, culminating in Beinn Udlamain (1011 m) and to the east is the undulating upland of the Gaick plateau, which reaches its highest point in Carn na Caim (941 m). The area is almost entirely underlain by Dalradian schists. During the last glacial maximum, ice nourished over Rannoch Moor moved northeastwards across Drumochter, in part guided by the troughs now occupied by Loch Ericht and Loch Garry. Granite erratics from Rannoch Moor are strewn across the area, even reaching the summit of Beinn Udlamain. Radiocarbon ages obtained for organic sediments at Loch Etteridge, 13 km northeast of Dalwhinnie, suggest that the last ice sheet retreated from the Drumochter area around 15.5 ka.

### The Loch Lomond Readvance

During the Loch Lomond Stade, Drumochter Pass was reoccupied by glaciers emanating from an icefield in the West Drumochter Hills (Fig. 10.11). At this time, glaciers flowing eastward into the pass advanced both northwards towards Dalwhinnie and southeast along Glen Garry, filling the pass with ice up to an altitude of ~650 m. Outside the limits of these glaciers, hillslopes are mantled by smooth solifluted till, and valley floors are occupied by outwash terraces above the present floodplain (Fig. 9.4). Inside these limits are multiple recessional moraines deposited as the Loch Lomond Readvance glaciers underwent oscillatory retreat, with brief readvances interrupting overall withdrawal of the ice margin (Fig. 5.15). On high ground the maximum altitude of the West Drumochter Icefield is represented by trimlines, where the lower limit of blockfields, debris-mantled slopes and relict, bouldery solifluction terraces coincides with a springline and the upper limit of glacial meltwater channels. This boundary is particularly clear at 750–800 m on the eastern slopes of Geal-charn (917 m) and A'Mharconaich (975 m), and on the southern slopes of Beinn Udlamain at about 850 m.

Along lower slopes flanking the northern shores of Loch Garry are low horizontal terraces at ~440 m altitude. These, and extensive terraces at the same altitude south of the loch, represent deltas deposited in a former glacial lake that was dammed at its northern end by the Glen Garry Glacier and at its southern end by another outlet glacier from the West Drumochter Icefield, raising the level of the loch about 20 m above its present elevation.

### Landslides and debris flows

Clearly visible from the A9 road near Dalnaspidal is a major rock-slope deformation on the steep west slope of Meall na Leitreach (NN 773634; 56.7460, -04.0079). Here an area of mountainside covering ~0.3 km² has moved downslope, but without developing the 'bulging' characteristic of some rock-slope deformations. Another rather curious rock-slope failure occurs at 650–750 m east of Drumochter Pass (NN 646755; 56.8511, -04.2218), where dislocated and tilted bedrock blocks separated by deep crevices are present on surprisingly gentle (~20°) slopes. Both landslides occurred after ice-sheet deglaciation but their ages are unknown.

Visible from the road on the northeast slope of An Torc (NN 622767; 56.8612, -04.2618) are deep gullies in the valley-side till deposits (Fig. 10.12). The four main gullies are up to 190 m long and 8 m deep, and extend downslope to feed debris cones composed of sediment that has been eroded, transported and redeposited by successive debris flows. At least 30,000 m³ of sediment has been removed from the slope during the Holocene, possibly since decline in forest cover after ~5.5 ka. The tracks of recent debris flows (parallel levées and terminal lobes of debris) are clearly evident on the debris cones. At least seven debris-flow events have occurred at this site during exceptional rainstorms over the past 40 years, making it possibly the most active site for repeated debris-flow activity in Scotland.

## Tinto Hill, Southern Uplands
### (NS 953344; 55.5919, -03.6629)
### Introduction

Tinto Hill (707 m) is a prominent mountain that straddles the Southern Uplands Fault. Most of the mountain is composed of pink felsite, a fine-grained igneous rock, and structurally it represents a *laccolith*, a dome-shaped igneous intrusion that has been unroofed through erosion of the overlying sandstones. The lower slopes are mantled by drift, but the summit above ~630 m is covered by frost-weathered debris comprising felsite clasts embedded in a matrix of silty sand; the latter is mixed with peaty soil, suggesting that soil, peat and vegetation cover were formerly more extensive, but have been removed by burning, overgrazing and erosion. The frost debris incorporates occasional erratics of andesite,

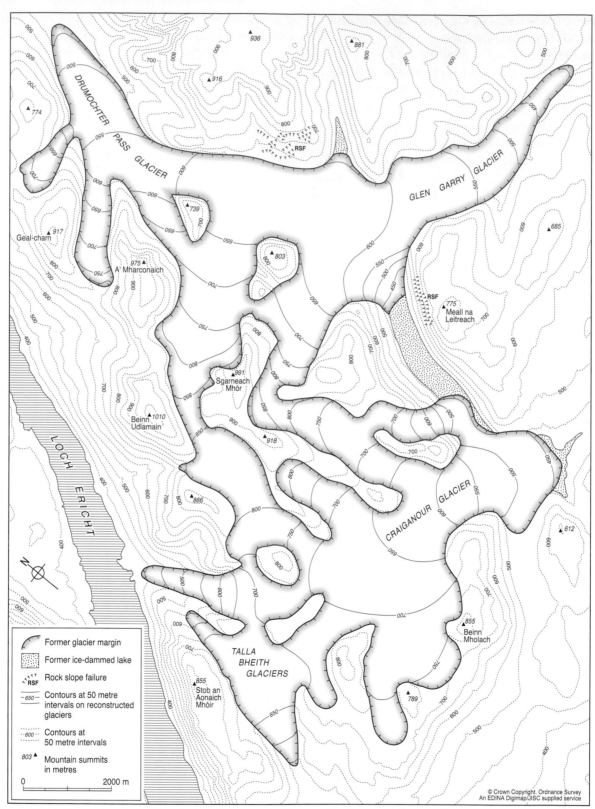

**Figure 10.11** Reconstruction of the glaciers that occupied Drumochter Pass and the west Drumochter Hills during the Loch Lomond Stade. Adapted from Benn, D.I. and Ballantyne, C.K. (2005) Palaeoclimatic reconstruction from Loch Lomond Readvance glaciers in the West Drumochter Hills, Scotland. *Journal of Quaternary Science*, 20, 577–592. © 2005 John Wiley and Sons Ltd.

**Figure 10.12** Gullies excavated in drift-covered hillslopes by repeated debris flows on the northeast slope of An Torc, Drumochter Pass. The eroded sediments have accumulated as debris cones. Levées and lobes of fresh debris represent recent debris-flow events.

sandstone and conglomerate that were deposited as the last ice sheet thinned. Erratics of Tinto felsite occur along the southern flanks of the Pentland Hills, indicating that the final ice movement across the mountain was northeastwards.

### Glacial meltwater channels and eskers

At the western end of Tinto Hill is a deep col channel, Howgate Mouth (NS 920342; 55.5894, -03.7151), where glacial meltwater flowing from south to north has incised a channel up to 20 m deep into bedrock. This channel has an 'up-and-down' long profile, implying that it was cut by a powerful subglacial meltwater stream that flowed northwards (uphill) under hydraulic pressure towards the highest point in the channel, then descended the far side. North of the notch, the channel links up with a series of eskers that trace the former route of meltwater flow northwards then eastwards at altitudes of 275–300 m. These in turn connect to a network of meltwater channels that extend over 3 km eastwards across the drift-covered terrain

north of Tinto from Lochlyoch Farm (NS 933361; 55.6068, -03.6953) to near Fallburn (NS 965376; 55.6209, -03.6451).

The configuration of the Tinto meltwater channel and esker system confirms that the final ice movement in this area was northeastwards. Tinto Hill therefore lay athwart the direction of ice movement, forming a barrier that forced meltwater under hydraulic pressure through the deep notch of Howgate Mouth, after which it followed multiple routeways along the northern foothill zone of Tinto towards the River Clyde; the channel of the Clyde probably hosted a subglacial river as the meltwater channel system developed.

### Active frost-sorted stone stripes

Tinto Hill supports the finest active frost-sorted patterned ground features in Scotland, in the form of stone stripes that have formed on patches of bare ground on its upper slopes. The best examples occur a short distance southeast of the summit at 610–660 m altitude (NS 954342; 55.5902, -03.6612), with subsidiary areas

**Figure 10.13** Active frost-sorted stone stripes on Tinto Hill.

of striped ground where a footpath crosses NE-facing slopes at 530 m on Maurice's Cleuch (NS 953351; 55.5982, -03.6632) and on a convex shoulder 700 m east of the summit at 540–560 m (NS 960344; 55.5921, -03.6518).

The stripes are aligned downslope and consist of alternating bands of fine and coarse felsite debris (Fig. 10.13); the coarse stripes are composed of slabby clasts 3–15 cm long, and the intervening fine stripes are covered by fine gravel. Trenches cut across the stripes show that the fine stripes consist of domed silt-rich soil, separated by gutters that are infilled by clasts to a depth of 15 cm. The main process responsible for sorting surface debris into fine and coarse stripes is the growth and collapse of needle ice during shallow freezing of the soil. Needle ice grows preferentially above patches of fine-textured soil, heaving up clasts. On thawing, the ice spicules bend, depositing clasts both downslope and at the margins of cells of finer-textured soil to form an embryonic stripe pattern. During more prolonged freezing events, ice lenses grow within the fine stripes, causing them to arch upwards, forming a 'ridge and furrow' micro-topography that enhances the stripe pattern. Over multiple freeze–thaw cycles, a regular spacing develops (Fig. 10.13) due to the establishment of a preferred 'wavelength' of differential frost heave that is related to the depth of soil freezing.

Disruption of part of the stripe pattern has demonstrated that mature frost-sorted stripes form within just 1–3 winters. Moreover, measurements of the downslope displacement of painted clasts on the stripes showed that these had moved 24–62 cm down a 23° slope over a single year as a result of recurrent episodes of needle-ice creep, the heaving of clasts normal to the slope on needle-ice crystals and their deposition farther downslope as the ice thawed. Such movement, however, is confined to a thin carpet of mobile surface debris.

# Further reading

Scores of books, memoirs and excursion guides and have been published on the geology and geomorphology of Scotland, and scientific papers on these topics in geography and geology journals run into thousands. The books and articles cited below represent the tip of an academic iceberg. They have been selected to provide readers who wish to pursue particular topics with appropriate source material, but are certainly not exhaustive.

## General Texts
Ballantyne, C.K. (2018) *Periglacial Geomorphology*. Chichester: Wiley-Blackwell, 452 pp.
Benn, D.I. and Evans, D.J.A. (2010) *Glaciers and Glaciation*. London: Hodder Education, 802 pp.
Bennett, M.R. and Glasser, N.F. (2009) *Glacial Geology: Ice Sheets and Landforms*. Chichester: Wiley-Blackwell, 385 pp.
Ehlers, J., Hughes, P. and Gibbard, P. (2016) *The Ice Age*. Chichester: Wiley-Blackwell, 548 pp.
Harvey, A.M. (2012) *Introducing Geomorphology: a Guide to Landforms and Processes*. Edinburgh: Dunedin Academic Press, 136 pp.
Huggett, R.J. (2017) *Fundamentals of Geomorphology*. London: Routledge, 543 pp.
Park, R.G. (2015) *Introducing Geology: a Guide to the World of Rocks*. Edinburgh: Dunedin Academic Press, 144 pp.

## Scottish geology and geomorphology
Ballantyne, C.K. and Harris, C. (1994) *The Periglaciation of Great Britain*. Cambridge: Cambridge University Press, 330 pp.
Gillen, C. (2013) *Geology and Landscapes of Scotland*. Edinburgh: Dunedin Academic Press, 246 pp.
Gordon, J.E. (2018) Shaping the landscape. In Kempe, N. and Wrightham, M. (eds) *Hostile Habitats: Scotland's Mountain Environment*. Aberdeen: Scottish Mountaineering Trust, pp. 56–87.
McKirdy, A., Gordon, J.E. and Crofts, R. (2007) *Land of Mountain and Flood: the Geology and Landforms of Scotland*. Edinburgh: Birlinn, 324 pp (reprinted 2017).
Trewin, N.H. (ed.) (2002) *The Geology of Scotland*. London: The Geological Society, 576 pp.
Upton, B. (2015) *Volcanoes and the Making of Scotland*. Edinburgh: Dunedin Academic Press, 248 pp.

## Chapter 2: Geological evolution of Scotland
Clarkson, E. and Upton, B. (2009) *Death of an Ocean: a Geological Borders Ballad*. Edinburgh: Dunedin Academic Press, 210 pp.
Goodenough, K. (2018) Geological foundations. In Kempe, N. and Wrightham, M. (eds) *Hostile Habitats: Scotland's Mountain Environment*. Aberdeen: Scottish Mountaineering Trust, pp. 24–53.
Gordon, J.E. (2010) The geological foundations and landscape evolution of Scotland. *Scottish Geographical Journal* 126, 41–62.
Gillen, C. (2013) *Geology and Landscapes of Scotland*. Edinburgh: Dunedin Academic Press, 246 pp.
Woodcock, N.H. and Strachan, R.A. (eds) (2012) *Geological History of Britain and Ireland*. 2nd edition. Chichester: Wiley-Blackwell, 454 pp.

## Chapter 3: Rocks, relief and the preglacial landscape
Hall, A.M. (1991) Pre-Quaternary landscape evolution in the Scottish Highlands. *Transactions of the Royal Society of Edinburgh: Earth Sciences* 82, 1–26.
Hall, A.M. and Bishop P. (2002) Scotland's denudational history: an integrated view of erosion and sedimentation at an uplifted passive margin. *Geological Society of London, Special Publications* 196, 271–290.

## Chapter 4: The Ice Age in Scotland
Ballantyne, C.K. and Small, D. (2019) The Last Scottish Ice Sheet. *Earth and Environmental Science Transactions of the Royal Society of Edinburgh* 110, 93–131.
Golledge, N.R. (2010) Glaciation of Scotland during the Younger Dryas stadial: a review. *Journal of Quaternary Science* 25, 550–566.

## Chapter 5: Glacial landforms

Ballantyne, C.K. (2017) The Dirc Mhór meltwater gorge. *Scottish Geographical Journal*, 133, 233–244.

Benn, D.I. (1992) The genesis and significance of 'hummocky moraine': evidence from the Isle of Skye, Scotland. *Quaternary Science Reviews* 11, 781–799.

Bickerdike, H.L., Evans, D.J.A., Stokes, C.R. and Ó Cofaigh, C. (2018) The glacial geomorphology of the Loch Lomond (Younger Dryas) Stadial in Britain: a review. *Journal of Quaternary Science* 33, 1–54.

Brook, M.S., Kirkbride, M.P. and Brock, B.W. (2004) Rock strength and development of glacial valley morphology in the Scottish Highlands and Northwest Iceland. *Geografiska Annaler* 86A, 225–234.

Evans, D.J.A., Hughes, A.L.C., Hansom, J.D. and Roberts, D.H. (2017) Glacifluvial landforms of Strathallan, Perthshire. *Scottish Geographical Journal* 133, 42–53.

Gordon, J.E. (2001) The corries of the Cairngorm Mountains. *Scottish Geographical Journal* 117, 49–62.

Rea, B.R. and Evans, D.J.A. (1996) Landscapes of areal scouring in NW Scotland. *Scottish Geographical Magazine* 112, 47–50.


## Chapter 6: Periglacial landforms

Ballantyne, C.K. (2019) After the ice: Lateglacial and Holocene landforms and landscape evolution in Scotland. *Earth and Environmental Science Transactions of the Royal Society of Edinburgh* 110, 133–171.

Goodfellow, B.W., Skelton, A., Martel, S.J., Stroeven, A.P., Jansson, K.N. and Hättestrand, C. (2014) Controls on tor formation, Cairngorm Mountains, Scotland. *Journal of Geophysical Research: Earth Surface* 119, 225–246.

Hopkinson, C. and Ballantyne, C.K. (2014) Age and origin of blockfields on Scottish mountains. *Scottish Geographical Journal* 130, 116–141.

Murton, J.B. and Ballantyne, C.K. (2017) Periglacial and permafrost ground models for Great Britain. *Geological Society, London, Engineering Group Special Publication*, 28, 501–597.


## Chapter 7: Landslides

Ballantyne, C.K. (2013) Lateglacial rock-slope failures in the Scottish Highlands. *Scottish Geographical Journal* 129, 67–84.

Ballantyne, C.K. Sandeman, G.F., Stone, J.O. and Wilson, P. (2014) Rock-slope failure following Late Pleistocene deglaciation on tectonically stable mountainous terrain. *Quaternary Science Reviews* 86, 144–157.

Curry, A.M. (2000) Holocene reworking of drift-mantled hillslopes in the Scottish Highlands. *Journal of Quaternary Science* 15, 529–541.

Hinchliffe, S. and Ballantyne, C.K. (2009) Talus structure and evolution on sandstone mountains in NW Scotland. *The Holocene* 19, 471–480.

Jarman, D. (2006) Large rock-slope failures in the Highlands of Scotland: characterisation, causes and spatial distribution. *Engineering Geology* 83, 161–182.

Jarman, D. & Ballantyne, C.K. (2002) Beinn Fhada, Kintail: a classic example of paraglacial rock-slope deformation. *Scottish Geographical Journal* 118, 59–68.

Luckman, B.H. (1992) Debris flows and snow avalanche landforms in the Lairig Ghru, Cairngorm Mountains, Scotland. *Geografiska Annaler* 74A, 109–121.


## Chapter 8: Aeolian landforms

Ballantyne, C.K. (1998) Aeolian deposits on a Scottish mountain summit: characteristics, provenance, history and significance. *Earth Surface Processes and Landforms* 23, 625–641.

Morrocco, S.M., Ballantyne, C.K., Spencer, J.Q.G. and Robinson, R.A.J. (2007) Age and significance of aeolian sediment reworking on high plateaux in the Scottish Highlands. *The Holocene* 17, 349–360.


## Chapter 9: Fluvial landforms

Addy, S., Soulsby, C., Hartley, A.J. and Tetzlaff, D. (2011) Characterisation of channel reach morphology and associated controls in deglaciated montane catchments in the Cairngorms, Scotland. *Geomorphology* 132, 176–186.

Ballantyne, C.K. (2019) After the ice: Lateglacial and Holocene landforms and landscape evolution in Scotland. *Earth and Environmental Science Transactions of the Royal Society of Edinburgh.* 110, 133–171.

Ballantyne, C.K. and Whittington, G.W. (1999) Late Holocene alluvial fan formation and floodplain incision, Central Grampian Highlands, Scotland. *Journal of Quaternary Science* 14, 651–671.

Foster, G.C., Chiverrell, R.C., Harvey, A.M., Dearing, J.A. and Dunsford, H. (2008) Catchment hydro-geomorphological responses to environmental change in the Southern Uplands of Scotland. *The Holocene* 18, 935–950.

Werritty, A. and McEwen, L.J. (1997) Fluvial geomorphology of Scotland. In Gregory, K.J. (ed.) *Fluvial Geomorphology of Great Britain*. Geological Conservation Review Series No. 13. London: Chapman and Hall, pp 21–32.

# Chapter 10: Key sites
## An Teallach and Torridon

Ballantyne, C.K. and Morrocco, S.M. (2006) The windblown sands of An Teallach. *Scottish Geographical Journal* 122, 149–159.

Ballantyne, C.K. and Stone, J.O. (2004) The Beinn Alligin rock avalanche, NW Scotland: cosmogenic [10]Be dating, interpretation and significance. *The Holocene* 14, 461–466.

Ballantyne, C.K. and Stone, J.O. (2009) Rock-slope failure at Baosbheinn, Wester Ross, NW Scotland: age and interpretation. *Scottish Journal of Geology* 45, 177–181.

Wilson, S.B. and Evans, D.J.A. (2000) Coire a'Cheud-chnoic, the 'hummocky moraine' of Glen Torridon. *Scottish Geographical Journa*, 116, 149–158.

## Trotternish and the Cuillin Hills, Skye

Ballantyne, C.K. (1989) The Loch Lomond Readvance on the Isle of Skye, Scotland: glacier reconstruction and palaeoclimatic implications. *Journal of Quaternary Science* 4, 95–108.

Ballantyne, C.K. (2007) Trotternish Escarpment, Isle of Skye, Highland. In Cooper, R.G. (ed.) *Mass Movements in Great Britain.* Geological Conservation Review Series No. 33. Peterborough: Joint Nature Conservation Committee, pp. 196–204.

Ballantyne, C.K and Lowe, J.J. (eds.) (2016) *The Quaternary of Skye: Field Guide.* London: Quaternary Research Association, 227 pp.

## The Cairngorms and Drumochter Pass

Benn, D.I. and Ballantyne, C.K. (2005) Palaeoclimatic inferences from reconstructed Loch Lomond Readvance glaciers, West Drumochter Hills, Scotland. *Journal of Quaternary Science* 20, 577–592.

Gordon, J.E. (2001) The corries of the Cairngorm Mountains. *Scottish Geographical Journal* 117, 49-62.

Hall, A.M. and Glasser, N.F. (2003) Reconstructing the basal thermal regime of an ice stream in a landscape of selective linear erosion: Glen Avon, Cairngorm Mountains, Scotland. *Boreas* 32, 191–207.

Hall, A.M. and Phillips, W.M. (2006) Glacial modification of granite tors in the Cairngorms, Scotland. *Journal of Quaternary Science* 21, 811–830.

Hall, A.M., Gillespie, M.R., Thomas, C.W. and Ebert, K. (2013) The Cairngorms – a pre glacial upland granite landscape. *Scottish Geographical Journal* 129, 2–14.

Rea, B.R. (1998) The Cairngorms – a landscape of selective linear erosion. *Scottish Geographical Magazine* 114, 124–129.

## Glen Roy and Glen Coe

Ballantyne, C.K. (2007) Coire Gabhail, Highland. In Cooper, R.G. (ed.) *Mass Movements in Great Britain.* Geological Conservation Review Series No. 33. Peterborough: Joint Nature Conservation Committee. pp.132–136.

Cornish, R. (2017) The gravel fans of upper Glen Roy, Lochaber, Scotland: their importance for understanding glacial, proglacial and glaciolacustrine dynamics during the Younger Dryas cold period in an Atlantic margin setting. *Proceedings of the Geologists' Association* 128, 83–109.

McEwen, L.J. (1994) Channel pattern adjustment and streampower variations on the middle River Coe, Western Grampian Highlands. *Catena* 21, 357–374.

Palmer, A.P. and Lowe, J.J. (2017) Dynamic landscape changes in Glen Roy and vicinity, west Highland Scotland, during the Younger Dryas and early Holocene: a synthesis. *Proceedings of the Geologists' Association* 128, 2–25.

Sissons, J.B. (2017) The Lateglacial lakes of Glens Roy, Spean and vicinity (Lochaber district, Scottish Highlands). *Proceedings of the Geologists' Association* 128, 32–41.

## Tinto Hill

Ballantyne, C.K. (2001) The sorted stone stripes of Tinto Hill. *Scottish Geographical Journal* 117, 313–324.

# Index of locations in Scotland

# Index of Scottish mountains and hills

# General Index

Page numbers in *italic* denote figures. Page numbers in **bold** denote tables.